ALIENS

JIM AL-KHALILI OBE is a British physicist, author and broadcaster.
He is currently Professor of Theoretical Physics and Chair in the
Public Engagement in Science at the University of Surrey. As well
as his research and writing, he has hosted many BBC TV and
radio   f the
inaugu   n.

ALIENS

Science asks: Is Anyone Out There?

Edited by Jim Al-Khalili

PROFILE BOOKS

First published in Great Britain in 2016 by
Profile Books Ltd
3 Holford Yard
Bevin Way
London WC1X 9HD
www.profilebooks.com

10 9 8 7 6 5 4 3 2 1

A CIP catalogue record for this book is available
from the British Library.

ISBN 978 1 78125 681 7
eISBN 978 1 78283 271 3

Typeset in Sabon by MacGuru Ltd
Printed and bound in Great Britain by
CPI Group (UK) Ltd, Croydon, CR0 4YY

Mixed Sources
Product group from well-managed
forests and other controlled sources
www.fsc.org Cert no. TT-COC-002227
© 1996 Forest Stewardship Council
FSC

Contents

Introduction: Where is Everybody?

Jim Al-Khalili

Enrico Fermi, the Italian-born American Nobel Prize winning physicist, made some of the most important contributions to twentieth-century science, but in 1950 he posed a very simple question that had nothing to do with his research in nuclear physics. It did, however, have very profound implications for anyone who is interested in the question of extraterrestrial life, as I assume you are since you are reading this book.

The story goes that the question in question came up during a lunchtime chat he was having with colleagues at the Los Alamos National Laboratory in New Mexico, sometime home of the Manhattan Project. They had been discussing the possibility that the Earth had been visited by aliens in flying saucers. The conversation was light-hearted and it doesn't appear that any of the scientists at that particular gathering actually believed in aliens. But Fermi asked a very simple question: 'Where is everybody?'

His point was that since the age of the Universe is so great and its size so vast, with almost half a trillion stars in the Milky Way alone, many of which would have their own planetary systems, then unless the Earth is astonishingly and unjustifiably special the Universe should be seething with life, including intelligent species advanced enough to have the knowledge and technology necessary for space travel. Surely then, he argued, we should have been visited by aliens at some point in our history. Indeed, maybe those accounts of flying saucer sightings that were being reported at the time were true. For Fermi it was obvious that, assuming our planet was not unique, not only should it be overwhelmingly likely that intelligent life exists elsewhere, but that there has been plenty of time for any alien civilisation with modest expansion ambitions and a sufficiently well-developed space travel technology to have colonised the entire Galaxy by now. So where are they all?

Fermi's conclusion was that the distances required for inter-stellar travel are so great that, owing to the theory of relativity's restriction that nothing can exceed the speed of light, no aliens would contemplate the ridiculously long journey to visit us. It seems he did not consider the notion that we should nevertheless be able to detect the existence of technologically advanced alien civilisations even if they never leave their home planet. After all, for the past hundred years or so, we have been announcing our presence to any listening aliens advanced enough and close enough to us (which means within 600 trillion miles of Earth, because that corresponds to 100 light years: the distance light will have travelled in 100 years). Ever since we invented radio and television, and more recently with the proliferation of satellite and mobile phone communication, we have been radiating our electromag-netic chatter into space. Any advanced enough aliens that are close enough to us and who happened to point their radio telescopes at our solar system would pick up faint signals betraying our presence.

Given that we have every reason to believe the laws of physics are the same everywhere in the Universe and that one of the easiest and most versatile means of transmitting information is by using electromagnetic waves, we should expect any advanced alien civili-sation to use this form of communication at some point during its technological development. And if it has, then some of these waves will inevitably leak out into space, spreading radially outwards through the Universe at the speed of light.

It is not surprising, therefore, that by the second half of the twentieth century, astronomers began to seriously consider the feasibility of listening out for such signals from space using their newly built radio telescopes. The search for extraterrestrial intelligence (or SETI) began with the pioneering efforts of one man, Frank Drake, who is probably most famous for coming up with a simple equation that bears his name and which includes all the factors he thought were necessary to provide an estimate of the likelihood that alien intelligence exists somewhere else in the Universe.

Today, SETI is the collective name for a number of projects around the world that have been conducted over the years to actively search for extraterrestrial signals. Following Frank Drake's initial projects, the SETI movement really took off, extending its search far beyond the solar system. The SETI Institute in California was set up in 1984 and several years later began Project Phoenix under the direction of astronomer Jill Tarter. Between 1995 and 2004, Project Phoenix used radio telescopes in Australia, the USA and Puerto Rico to look at hundreds of Sun-like stars within a couple of hundred light years of Earth. So far, they have heard nothing. But the project has produced a valuable source of information for research into possible alien life. Today, hunting for extrasolar planets (those orbiting stars other than our Sun) is one of the hottest areas of scientific research, and with bigger and more powerful radio telescopes at their disposal, astronomers are discovering new, potentially habitable star systems on a regular basis. Indeed, it seems that barely a month goes by without news of new Earth-like planets that have the potential to harbour life.

The announcement in 2015 that SETI will invest $100 million in the quest to discover intelligent life elsewhere in the Universe has captured the public's imagination around the world. The physicist Stephen Hawking spoke for many when he commented: 'It's time to commit to finding the answer, to search for life beyond Earth. It is important for us to know if we are alone in the dark.'

Other academic research, however, has in recent years focused not on searching for radio signals sent by intelligent life forms, but for the planets and moons themselves that might host them. Closest to home, we have extended the search beyond Mars to the moons of Jupiter and Saturn. And then there are the extrasolar planets. Much excitement currently surrounds the James Webb Space Telescope, due to be launched in 2018, that will represent the next generation of space telescopes and will be the first that is truly capable of detecting the biosignatures of alien life.

Of course, an alien planet being suitable for life is one thing, but the really big unknown is this: given the right conditions, how likely is it that life could evolve elsewhere? To answer is that we need to understand how life began on Earth. If we are indeed alone in the vastness of the cosmos, then we need to understand why we are so special. Why would the Universe be apparently so finely tuned for life to exist, yet harbour it in just one isolated corner?

One way of thinking about this is to ask yourself how come *you* exist? What were the chances that your parents would meet and produce you? Indeed, what were the chances of their parents meeting, and so on all the way back? We are each of us the culmination of a long and highly unlikely chain of events leading back to the origin of life itself. Break any one of the links in that chain and you would not be here to ask the question in the first place. Maybe our existence is really no more remarkable than the lottery winner contemplating his or her good fortune: had that sequence of numbers not come up, then someone else would have won and they would also reflect on the improbable odds of their win.

What life on Earth can tell us about alien existence elsewhere in the Galaxy is limited by the fact that we have a statistical sample of just one. Our own example tells us nothing about the likelihood of life elsewhere, or what it would look like if it did exist. Could there be advanced alien civilisations out there or would they only be in the form of simple, single-celled microbes? If we can't begin to address that issue, how will we even know where to look?

Most profound of all of course is what it would mean for us if we did find them? We've come a long way since the days of flying saucer sightings and scientists these days take the whole search for extraterrestrial life very seriously. In this book I've handpicked a quite remarkable team of scientists and thinkers, world leaders in their fields who will cover all aspects of the subject.

So before you plunge in, dear reader, allow me to introduce my 'Team Aliens'. You will find that each and every one of them offers his or her own unique perspective on the subject.

Leading us gently out on our cosmic journey is the Astronomer Royal, cosmologist Martin Rees, who in Chapter 1 speculates on our place in the Universe, giving a brief history of mankind's ideas on the subject and projecting forward into the distant future to consider whether one day we ourselves will be the 'aliens', exploring space and colonising the Galaxy.

In Chapter 2, astrobiologist Lewis Dartnell asks a question that Enrico Fermi may well have contemplated: if advanced spacefaring aliens are out there, what would motivate them to visit us? He explores whether an alien invasion would be the end of mankind as we know it or a meeting of mutually curious peaceful civilisations.

In Chapter 3, science broadcaster Dallas Campbell gives an entertaining historical overview of our obsession with aliens and alien sightings ever since the celebrated Kenneth Arnold flying saucers case of 1947. If you want to get the *real* inside story on the possibility of alien life, it's worth getting the conspiracy theories and wackier mythology out of your system before moving on to the serious science – which you can easily do with Dallas's vivid account of Roswell, Area 51, the 'Men in Black', and alien abductions.

In Chapter 4, cognitive neuroscience and artificial intelligence expert, Anil Seth, explores how alien intelligence might differ from our own by studying the most alien intelligence we can find here on Earth: the octopus. As he says, there is no need to travel to a distant planet to encounter alien intelligence. You can find that 'otherworldliness' here on Earth by examining how the mind of an octopus works.

Chris French is a psychologist and professor of paranormal belief and experiences, and in particular the belief in conspiracy theories and false memories. In Chapter 5, he argues that there are millions of people around the world who believe that speculation on the existence of alien life is a waste of time – because there is already convincing evidence to show that aliens not only exist, but have already made contact with us. According to French, however, there are well-established psychological phenomena which may explain such 'close encounters'.

In Chapters 6, 7 and 8, we begin our quest in earnest. NASA astrobiologist Chris McKay begins, in Chapter 6, by asking what the ingredients for life elsewhere might be. In some ways, you might think the answer is obvious: surely all life needs energy – that's a given. But what about water? And what of the various elements, such as carbon and oxygen, and the molecular building blocks they would have to form? How vital are they and are we being imaginative when we think about the limits of life?

McKay hands over to space scientist Monica Grady and planetary geologist Louisa Preston as we head off into the solar system. Our first port of call is, naturally, our nearest neighbour, Mars. In Chapter 7, Grady rightly begins, 'In any book about aliens, there has to be a chapter about Mars'; she goes on to explore how Mars differs from Earth and whether there was a time, billions of years ago, when it might have been teeming with life – instead of the barren wasteland it appears to be today. In Chapter 8, Louisa Preston takes us to the outer planets of the solar system – the gas giants of Jupiter and Saturn in particular – and asks whether their giant moons – Europa, Enceladus and Titan – while possessing far more hostile environments than Earth, might in fact be home to *some* form of hardy microbial life.

Having explored what aliens might look like in reality, mathematician Ian Stewart introduces us to some of the more imaginative manifestations of alien life. I have known Ian for many years and was aware of his great love of science fiction – he has a remarkable collection of over 8000 SF books – and so I invited him to explore aliens in science fiction writing in Chapter 9. From H. G. Wells and A. E. van Vogt to Arthur C. Clarke, Larry Niven, Stephen Baxter and, my personal favourite, Robert Heinlein, if you thought all fictional aliens were little green men with bug eyes and ray guns, then have a look at what these writers have dreamt up. Stewart also brings a sceptical eye to the scientific principles involved in inventing plausible aliens, and the extent to which SF writers have toed the line.

We are now well into our stride and come to one of the thorniest issues in the book. You see, in order to assess the likelihood of alien life somewhere else in the Universe, we really need to understand how special life is, and how and why it emerged on Earth. Chapters 10, 11 and 12 explore the science of life itself. First, chemist Andrea Sella takes us back to basics. Ultimately, all biology must boil down to chemistry, so are there chemical reactions that can drive a system towards complexity – from inanimate matter to something that is able to maintain a highly organised state? He then hands over to biochemist Nick Lane, who examines the origins of life on Earth in Chapter 11. If you thought it was simply a matter of mixing all the chemical ingredients together under the right conditions in some warm pond nearly 4 billion years ago, then you're way behind the times. Science may not have solved the mystery of the origin of life, but it has made great strides in recent years. Lane first defines what it means for something to be 'alive' and then explores some possible ways that chemistry might have become biology.

In Chapter 12, my long-time colleague and collaborator, molecular geneticist Johnjoe McFadden, adds a novel ingredient to the mix. He argues that the sheer improbability of life emerging spontaneously on Earth almost as soon as the conditions were right for it cannot be explained away so easily. He postulates that quantum mechanics, that strangely counter-intuitive theory of the subatomic world, may have played a crucial role in speeding things up.

Theoretical physicist Paul Davies has written extensively on the question of whether life exists elsewhere in the Universe. Among his many activities he has the intriguing role of acting chairman of SETI's 'Post-Detection Science and Technology Task Group', whose job it is 'to be available to be called on at any time to advise and consult on questions stemming from the discovery of a putative signal of extraterrestrial intelligent origin'. I think this means he is the person who announces the news to the world if and when we discover aliens. In Chapter 13, he examines the likelihood of alien life from a broader cosmological perspective and ponders why so many distinguished scientists are convinced that life must exist beyond our planet.

This book is nothing if not balanced, and in Chapter 14, zoologist Matthew Cobb provides a sobering counter-argument to the optimism of the previous few chapters. He claims that the emergence of life on Earth, and complex multicellular (and intelligent) life in particular, was so incredibly unlikely that his answer to the Fermi question I posed at the beginning could be summed up as another question: Why should we even expect there to be anyone else?

In Chapter 15, geneticist and broadcaster Adam Rutherford explores the way film-makers have portrayed aliens in the movies. He takes us on an entertaining and rich digression through a century of cinema, from the brilliantly plausible to the just plain awful, the common thread being that almost all of them have given us a vision of aliens that are remarkably like us – which is almost certainly wrong.

Finally, we are ready to explore the vast expanse of space. The common theme of the next four chapters is that their authors, all world-class scientists, search for extraterrestrial life for a living. Astrobiologist Nathalie Cabrol is director of the Carl Sagan

Center and has been a leading SETI researcher for almost two decades. In Chapter 16, she gives us an insider's perspective on the search for extraterrestrials (past, present and future). Then, in Chapter 17, MIT astronomer Sara Seager reviews what will be possible with the new James Webb Space Telescope, and updates Drake's famous equation to give us a way of calculating the likelihood of alien life using some of the most recent advances in our understanding.

Chapter 18, by astrophysicist Giovanna Tinetti, describes how we are now able to use a technique known as spectroscopy, which can do more than just detect distant Earth-like exoplanets. In early 2016, she was one of the authors of a paper that reported the first direct identification and measurement of gases in the atmosphere of an exoplanet – one twice the size of Earth, orbiting a yellow dwarf star called Copernicus, 41 light years away in the constellation of Cancer. Discovering what the atmosphere of a distant planet is made up of is a fantastic way of looking for the telltale signs of life there. For example, if we find oxygen, water vapour, or complex organic compounds, then I for one would be very excited.

Last but most definitely not least, Chapter 19 is the contribution of the current director of SETI, the astronomer Seth Shostak, who stresses just how ingenious, creative and resourceful we will need to be in our search for life elsewhere.

All these essays, and the work of the pioneering scientists and writers that they are based on, reflect the fact that today, in the second decade of the twenty-first century, we are only just beginning our adventure, seeking answers to the most fundamental questions of existence: What is life? Are we unique? And what is our place in the Universe?

The search for aliens is a subject that has a reputation for being light-hearted, and even silly at times – beset with conspiracy theories and little green men – but, actually, thinking about extraterrestrials has led us to ask, and even begin to answer, some of the most profound questions about our own existence. What has changed in recent years is that deep questions such as these are no longer just the preserve of theologians and philosophers – serious scientists have joined in too. What's more, we are actively doing something about it. This collection of essays will help you make up your own mind. I'm sure you'll enjoy them.

Stop press:

Oh, and before we begin our journey, I have to mention a recent exciting discovery. In a way, I guess it emphasises just how thrilling and fast-moving the field of astronomical research is at the moment.

At a distance of 4.25 light years from our solar system, Proxima Centauri is our closest stellar neighbour. It is a small, moderately active red dwarf star and, with a surface temperature of under 3,000 degrees Celsius, it is considerably cooler than our sun. On 24 August 2016, the European Southern Observatory announced the discovery of an Earth-sized planet, dubbed Proxima b, in a tight orbit around it – with an orbital radius just 5 per cent that of the earth's around the sun, making its year just eleven earth days long. What is exciting is that this rocky planet, with an estimated mass at least 1.3 times that of the Earth, would have a temperature estimated to be within the range where liquid water could exist on its surface, meaning that it sits in the habitable zone of its star.

Moreover, at just over four light years away, there is even hope that we might one day be able to visit Proxima b to check it out for ourselves. In fact, at the time of writing, a planned unmanned mission to Proxima, named Project Starshot, and which would involve a fleet of micro-spacecraft, powered by laser beams, is on the cards. Travelling at a fifth of the speed of light, these craft could reach Proxima b in twenty years and beam back information on whether there is life there. Who knows what we might find.

Jim Al-Khalili, *26 August, 2016*

1

Aliens and Us: Could Post-humans Spread through the Galaxy?

Martin Rees

Extraterrestrial life and alien intelligence have always been fascinating topics on the speculative fringe of science. But in the last decade or two, serious advances on several fronts have generated wider interest in these subjects. They have become almost 'mainstream' – vibrant frontiers of science.

The study of planets orbiting other stars, known as exoplanets, began only twenty years ago. Now, we can confidently assert that there are billions of them in our Galaxy. There has also been progress in understanding the origin of life. It's been clear for decades that the transition from complex chemistry to the first entities that could be described as 'living' poses one of the crucial problems in the whole of science. But until recently, people shied away from it, regarding it as neither timely nor tractable. In contrast, numerous distinguished scientists are now committed to this challenge.

Advances in computational power and robotics have led to growing interest in the possibility that 'artificial intelligence' (AI) could in the coming decades achieve (and exceed) human capabilities. This has stimulated discussions of the nature of consciousness, and further speculation by ethicists and philosophers on what forms of inorganic intelligence might be created by us – or might already exist in the cosmos – and how humans might relate to them.

Some history

Speculations on 'the plurality of inhabited worlds' date back to antiquity. From the seventeenth to the nineteenth century, it was widely suspected that the other planets of our solar system were inhabited. The astronomer William Herschel even thought the Sun might be inhabited. The arguments were often more theological than scientific. Eminent nineteenth-century thinkers argued that life must pervade the cosmos because, otherwise, such vast domains of space would seem such a waste of the Creator's efforts. An amusing critique of such ideas is given by the co-developer of natural selection theory, Alfred Russel Wallace, in his hugely impressive book *Man's Place in the Universe*. Wallace is specially scathing about the physicist David Brewster (remembered for the 'Brewster angle' in optics), who conjectured on such grounds that even the Moon must be inhabited. Brewster argued that had the Moon 'been destined to be merely a lamp to our Earth, there was no occasion to variegate its surface with lofty mountains and extinct volcanoes … It would have been a better lamp had it been a smooth piece of lime or of chalk.'

By the end of the nineteenth century, so convinced were many astronomers that life existed on other planets in our solar system that a prize of 100,000 francs was offered to the first person to make contact with them. And the prize specifically excluded contact with Martians – that was considered far too easy!

The space age brought sobering news. Venus, a cloudy planet that promised a lush tropical swamp-world, turned out to be a crushing, caustic hell-hole. Mercury was a pockmarked blistering rock. And NASA's Curiosity probe (and its predecessors) showed that Mars, though the most Earth-like body in the solar system, was actually a frigid desert with a very thin atmosphere.

There may be creatures swimming under the ice of Jupiter's moon Europa, or Saturn's moon Enceladus, but nobody can be optimistic. Certainly there's no expectation of advanced life anywhere in the solar system away from the Earth.

However, once we look beyond our own solar system, further than any physical probe can reach today – our prospects seem brighter. It's now clear that most stars are orbited by their own planets, something that Giordano Bruno speculated about in the sixteenth century. From the 1940s onward, astronomers suspected he was correct: the earlier idea that our solar system formed from a filament torn out of the Sun by the tidal pull of a close-passing star (which would have implied that planetary systems were rare) had by then been discredited. But it wasn't until the late 1990s that evidence for exoplanets started to emerge. There's huge variety among these systems, but there are probably around a billion planets in the Milky Way that are 'Earth-like' in the sense that they are about the size of the Earth and at a distance from their parent star such that water can exist, neither boiling away nor staying permanently frozen.

These planets would be 'habitable'. But of course that doesn't mean that they are inhabited – indeed we can't yet exclude the possibility that life may have begun with a chance event so unlikely that it might only have occurred once in the entire Galaxy. But it's also possible that life is destined to emerge once the right conditions appear for it to do so. We just don't know. Nor do we know if the DNA/RNA chemistry of terrestrial life is the only possibility, or just one chemical basis among many options that could be realised elsewhere.

But this key issue may soon be clarified. The origin of life is now attracting stronger interest: it's no longer deemed to be one of those problems (like consciousness) which, though manifestly important, is relegated to the 'too difficult box'.

Almost half a millennium ago, Giordano Bruno famously went even further, and conjectured that on some of those planets there might be other creatures 'as magnificent as those upon our human Earth'. Will he one day be proved right: if simple life exists, is it likely to evolve into a sophisticated, conscious being like ourselves? Even if some form of microbial life emerges fairly readily, intelligence may not necessarily follow, and may depend on a vast number of – mainly unknown – factors. The course of evolution on Earth was influenced by phases of glaciation, the Earth's tectonic history, asteroid impacts, and so forth. Several authors have speculated about possible 'bottlenecks' – key stages in evolution that are hard to transit. Perhaps the transition to multicellular life is one of these. (The fact that simple life on Earth seems to have emerged quite quickly, whereas even the most basic multicellular organisms took nearly 3 billion years, suggests that there may be severe barriers to the emergence of any complex life.) Or the 'bottleneck' could come later. Even in a complex biosphere the emergence of intelligence isn't guaranteed. Had the dinosaurs not been wiped out, leaving an evolutionary vacuum for our mammalian ancestors to evolve into, it's impossible to know whether some other intelligent creature would have emerged in our place.

Perhaps, more ominously, there could be a 'bottleneck' at our own present evolutionary stage – the stage when intelligent life develops powerful technology. If so, the long-term prognosis for 'Earth-sourced' life depends on whether humans survive this critical evolutionary phase. This does not mean that the Earth has to avoid a disaster – only that, before it happens, some humans or advanced artefacts have spread beyond their home planet.

In considering the possibilities of life elsewhere, we should surely be open-minded about where it might emerge and what forms it could take – and devote some thought to non-Earth-like life in non-Earth-like locations. But it plainly makes sense to start with what we know and to deploy all available techniques to discover whether any exoplanet atmospheres display evidence for a biosphere. Clues will surely come, in the next decade or two, from high-resolution spectra using the James Webb Space

Telescope and the next generation of 30-metre ground-based telescopes which will come online in the 2020s. To optimise the prospects, we shall need beforehand to have scanned the whole sky to identify the nearest Earth-like planets. Even these next-generation telescopes will have a hard job separating out the spectrum of the planet's atmosphere from the spectrum of the hugely brighter central star.

Conjectures about advanced or intelligent life are of course far more shaky than those about simple life. I would argue however, that they suggest two things about the entities that SETI searches could reveal:

1. They will not be 'organic' or biological.
2. They will not remain on the planet where their biological precursors lived.

The far future of Earth-sourced intelligence

We have already begun to explore our solar system. By the end of the century, we should have mapped every planet, moon and asteroid. The next development will be large-scale robotic fabricators, capable of building large structures in space – something which will prove a more effective use of resources mined from asteroids or the Moon than bringing them back to Earth. The successors of the Hubble Space Telescope, with vast, leaf-thin mirrors assembled under zero gravity, will allow us to look deeper into the cosmos than ever before.

But what role will humans play? There's no denying that NASA's Curiosity, now trundling across a giant Martian crater, may miss startling discoveries that no human geologist could overlook. But robotic techniques are advancing fast, allowing ever more sophisticated unmanned probes – and, later in the century, robotic 'explorers' endowed with human level intelligence. The practical case for manned space flight gets ever weaker with each advance in robotics and miniaturisation. If some people now living one day walk on Mars (as I hope they will) it will be as an adventure, and as a step towards the stars.

When conducted by Western governments, manned space flight is likely to remain hugely expensive. Costs can be cut if the astronauts are prepared to face higher risks – and I think those who venture beyond the Moon will be adventurers prepared to accept these risks. But don't ever expect mass emigration – nowhere in our solar system is as sympathetic to life as even the midwinter Antarctic icepack or the highest mountain peaks. We cannot look to space for an escape from Earth's problems.

Despite this, within a century or two, it is possible that small groups of pioneers will have established a habitat where they can live independently from Earth. We may, on ethical grounds, wish to restrain the use of genetic modification and cyborg enhancement here on Earth. But we should surely expect (and welcome) the efforts of these pioneers to seize on all new technologies that allow their descendants to thrive in an alien environment. Within a few centuries they will have diverged into a new species: the post-humans – quite different from those who remain on Earth. Eventually, there may be a transition to fully inorganic intelligences.

The most crucial impediment to routine space flight in Earth's orbit, and still more for those venturing further, stems from the intrinsic inefficiency of chemical fuel, and the consequent requirement to carry a weight of fuel far exceeding that of the payload. So long as we are dependent on chemical fuels, interplanetary travel will remain a challenge. (It's interesting to note, incidentally, that this is a generic constraint, based on fundamental chemistry, on any organic intelligence that had evolved on another planet. If a planet's gravity is strong enough to retain an atmosphere, at a temperature where water doesn't freeze and metabolic reactions aren't too slow, the energy required to lift a molecule from it will require more than one molecule of chemical fuel.)

Nuclear energy (or, more futuristically, matter–antimatter annihilation) could be a transformative fuel. Even with this, however, the time it would take to travel further than our nearby stars would be longer than a human lifespan. Interstellar travel will only be a realistic possibility for post-humans. Whether they will have left their biological trappings behind and become silicon-based, or whether they will be organic creatures who have found a way to avoid or indefinitely delay the natural processes of aging and death, remains to be seen.

Few would doubt that machines will eventually take on – and even outdo – many of the capabilities that we currently think of as uniquely human. The question is when. Will it be a process spread over hundreds of years, or a few short decades? Our latest phase of evolution is moving far more rapidly than those that have gone before: compared to the billions of years of Darwinian evolution that have got us to the current point – and to the aeons of cosmic time that lie ahead – it will happen in an instant. The products of technological evolution may value our current intellect in much the same way that we would perceive a moth's.

So humans are not the apex of evolution: if we trigger the transition to silicon-based, potentially immortal entities, our role may still be of special cosmic significance – and that should be a consolation to us. As post-humans journey into space and transcend the limitations of biological entities, our legacy will be their influence in the wider cosmos. Just as we are able to see around us the marks of past civilisations, we will leave our own archaeological traces for the inhabitants of the far future to discover. Later, larger and more sophisticated cultures, millennia hence, may retain the imprint of our thoughts and beliefs in much the same way that our own bodies contain vestiges of earlier evolutionary stages.

A further question is whether such beings would be properly 'conscious', or whether that is a property arising solely from the wet, organic brains of humans and perhaps a few animals. Regardless of their capabilities, will robots become self-aware, or have the kind of vivid interior life that is a feature of our own minds? The answer to this question crucially affects how we react to the far-future scenario I've sketched. If the machines are zombies, the post-human future would seem bleak. But if they are conscious, we should surely welcome the prospect of their future hegemony.

Many thinkers today acknowledge that it seems likely that machines will come to overtake us on Earth, perhaps before we are able to establish any kind of independent satellite community of humans elsewhere in space. Abstract thinking by biological brains has underpinned the emergence of all culture and science. But this activity – spanning tens of millennia at most – will be a brief precursor to the more powerful intellects of the inorganic post-human era. There are chemical and metabolic limits – which we may already have reached – to what the 'wet' organic brain is capable of. Silicon-based (or even quantum) computers have no such limitations, and their future development could be as dramatic and significant as the evolution from monocellular organisms to humans.

It is also the case that, while we are tied to this planet, having evolved in tandem with it, AI has no such constraints, and it may be that it is in interplanetary and interstellar space that they will have scope to develop to their fullest capacity.

Our brains have changed little since we first emerged as a species, and it is astonishing that they have equipped us to understand not only the challenges of survival on the African savannah but also the counter-intuitive concepts inherent in quantum mechanics and the cosmos. Nonetheless, some key features of reality may be beyond our conceptual grasp. As scientific frontiers advance, answers to many current mysteries will surely come into focus. But there may be aspects of the Universe, which will shape our long-term destiny, that are simply beyond our understanding. We may have to await a post-human intelligence that may structure its consciousness in an entirely different way before these mysteries are revealed.

Some thoughts on SETI: prospects and techniques

The scenarios I've just described would have the consequence – a boost to human self-esteem! – that even if life had originated only on Earth, it would not remain a trivial feature of the cosmos: humans may be closer to the beginning than to the end of a process whereby ever more complex intelligence spreads through the Galaxy. But of course there would in that case be no 'ET' at the present time.

Suppose, however, that there are many other planets where life began; and suppose that on some of them Darwinian evolution followed a similar track. Even then, it's highly unlikely that the key stages would be synchronised. If the emergence of intelligence and technology on a planet lags significantly behind what has happened on Earth (because the planet is younger, or because the 'bottlenecks' have taken longer to negotiate there than here) then that planet would plainly reveal no evidence of ET. But life on a planet around a star older than the Sun could have had a head start of a billion years or more. Thus it may already have evolved much of the way along the futuristic scenarios outlined in the last section.

One generic feature of these scenarios is that 'organic' human-level intelligence is just a brief interlude before the machines take over. The history of human technological civilisation is measured in millennia (at most) – and it may be only one or two more centuries before humans are overtaken or transcended by inorganic intelligence, which will then persist, continuing to evolve, for billions of years. This suggests that if we were to detect ET, it would be far more likely to be inorganic: we would be most unlikely to 'catch' alien intelligence in the brief sliver of time when it was still in organic form.

SETI programmes looking for alien life are surely worthwhile, despite the heavy odds against success, because the stakes are so high. Breakthrough Listen – a major ten-year commitment by the Russian investor Yuri Milner to employ and develop technology to scan the sky more thoroughly than ever before – can only be applauded.

The radio telescopes Milner's grant will secure time on will be used to search for non-natural radio transmissions from nearby and distant stars, from the plane of the Milky Way, from our galactic centre, and from nearby galaxies. They will search over a wide range of radio and microwave frequencies using advanced signal processing, seeking some transmission that is manifestly artificial. But even if the search succeeded (and few of us would bet more than 1 per cent on this), it would still in my view be unlikely that the 'signal' would be a decodable message. It would more likely represent a by-product (or even a malfunction) of some supercomplex machine far beyond our comprehension that could trace its lineage back to alien organic beings (which might still exist on their home planet, or might long ago have died out). The only type of intelligence whose messages we could decode would be the (perhaps small) subset that used a technology attuned to our own parochial concepts.

Even if intelligence were widespread in the cosmos, we may only ever recognise a small and atypical fraction of it. Some 'brains' may package reality in a fashion that we can't conceive. Others could be living contemplative energy-conserving lives, doing nothing to reveal their presence. We should look first to planets like our own, orbiting long-lived stars, even if the worlds conjured up in science fiction provide more exciting visions of where life might be found. In particular, the habit of referring to ET as an 'alien civilisation' may be too restrictive. A 'civilisation' connotes a society of individuals: in contrast, ET might be a single integrated intelligence. Even if signals were being transmitted, we may not recognise them as artificial because we may not know how to decode them. A radio engineer familiar only with amplitude modulation might have a hard time decoding modern wireless communications.

Perhaps the Galaxy already teems with advanced life, and our descendants will 'plug in' to a galactic community – as rather 'junior members'. Alternatively, we may be alone after all – the lucky recipients of a cosmic habitat that seems 'tuned' for life. If this is the case, we can be less modest: our tiny planet – this pale blue dot floating in space – could be the most important place in the entire cosmos: the place where life emerged and, as the drive towards consciousness and complexity continues, took off and spread across the Universe.

Finally, there are two familiar maxims that pertain to this quest. First 'extraordinary claims will require extraordinary evidence' and second 'absence of evidence isn't evidence of absence'.

CLOSE ENCOUNTERS

2

(Un)welcome Visitors: Why Aliens Might Visit Us

Lewis Dartnell

As an astrobiologist I spend a lot of my time working in the lab with samples from some of the most extreme places on Earth, investigating how life might survive on other worlds in our solar system and what signs of their existence we could detect. If there is biology beyond the Earth, the vast majority of life in the Galaxy will be microbial – hardy single-celled life forms that tolerate a much greater range of conditions than more complex organisms can. Some of the other authors in this book discuss the reasons why intelligent life may well be vanishingly rare in the Galaxy, and to be honest, my own point of view is pretty pessimistic too. Don't get me wrong – if the Earth received an alien tweet tomorrow, or some other text message beamed at us by radio or laser pulse, then I'd be absolutely thrilled. So far, though, we've seen no convincing evidence of other civilisations among the stars in our skies.

But let's say, just for the sake of argument, that there are one or more star-faring alien civilisations in the Milky Way. We're all familiar with Hollywood's darker depictions of what aliens might do when they come to the Earth: zapping the White House, harvesting humanity for food like a herd of cattle, or sucking our oceans dry. These scenarios make great films, but don't really stand up to rational scrutiny. So let's run through a thought experiment on what reasons aliens might possibly have to visit the Earth, not because I reckon we need to ready our defences or assemble a welcoming party, but because I think considering these possibilities is a great way of exploring many of the core themes of the science of astrobiology.

Aliens come to Earth to enslave humanity or for breeding partners. Alien races enslaving each other is a common trope of many science fiction Universes. While enslavement of defeated enemies or other vulnerable populations has regrettably been a common feature of our history on Earth, it's hard to see why a species with the capability of voyaging between the stars, and therefore having already demonstrated the mastery of a highly advanced level of machinery and of marshalling energy resources, would have any need for slaves. Constructing robots, or other forms of automation or mechanisation, would be a far more effective solution for labour – people are feeble in comparison, harder to fix, and need to be fed. Likewise, the idea of an alien species needing humans for breeding doesn't really stand up to scrutiny. The act of sexual reproduction, on a genetic level, involves the combination of DNA from two individuals. So on the most fundamental level, for an alien race to be compatible with us, they would need not only to use the same polymer, deoxyribonucleic acid, as the storage molecule for their genetic information, but also to use the same four 'letters' for their genetic alphabet (and not other purine and pyrimidine bases that exist in chemistry), and the same coding system for translating those sequences of genetic letters into proteins, and the same organisational structure of the DNA strands into chromosomes, and so on. There is a lot of ongoing research on whether extra-terrestrial life is likely to use DNA, or what molecular alternatives there might be, but it is a huge stretch to expect alien life to be that similar to human genetics. Humans can't even interbreed with our closest evolutionary relatives on Earth, the chimpanzees (indeed, this is the basis of the definition for different species – two organisms which are not able to reproduce fertile offspring), and so it is overwhelmingly improbable that an alien life form from a completely different evolutionary lineage would be compatible.

Aliens come to Earth to harvest us for food. If aliens wouldn't be bothered about enslaving or breeding with us, might they simply be coming to Earth for a drive-by meal? The question of whether an alien biochemistry would be able to digest us as food actually comes down to some very fundamental features of the molecules of life. Other chapters in this book describe the core molecular basis of all terrestrial organisms. Our cells are made up of various organic molecules: proteins (polymers of amino acids), nucleic acids DNA and RNA (polymers of bases and sugars), and membranes of phospholipids. And so for making more cells for reproduction, growth and repair of our bodies we need a source of these simple building blocks. We eat other animals or plants and our digestive system breaks them down into their component amino acids, sugars and fatty acids, which we then use as the building blocks for ourselves. So in order to derive any useful nutrition from eating a human, an alien monster would need to be based on very similar biochemistry, and thus have the enzymes needed for processing the molecules we are built from. A whole variety of amino acids, sugars and fatty molecules are actually found in certain meteorites, having been produced by astrochemistry in outer space, and so maybe extraterrestrial life would be based on the same basic building blocks as us. But there's another, very interesting subtlety here. Simple organic molecules like amino acids and sugars can exist in two different forms, mirror images of each other (in the same way your two hands are similar shapes but can't be placed exactly one on top of the other). These two versions are known as enantiomers, and it turns out that all life on Earth uses only left-handed amino acids and right-handed sugars, whereas non-living chemistry produces even mixtures of both kinds. So if we do find traces of amino acids on Mars, one very good way of telling whether these organic molecules are the relics of ancient Martian life or are just the product of astrochemistry would be to check if they are mostly left- or right-handed forms, or just an even mixture. The most exciting discovery would be to detect traces of ancient bacteria on Mars and to find that they employ the opposite forms of organic molecules to us: right-handed amino acids or left-handed sugars, because then we would know for sure that this life was definitely extraterrestrial and not merely contamination from Earth. So here's a fascinating thought: alien invaders could be based on exactly the same organic molecules (amino acids, sugars, etc.), but they still wouldn't gain any nutrition from eating us as the origins of life on their own planet settled on the opposite enantiomers. We'd be mirror images of each other, on a molecular level.

Aliens come to Earth to suck our oceans dry. If alien marauders would need to have an essentially identical biochemistry to bother culling us for food, maybe they come to Earth to harvest some other vital substance. All life on Earth is water-based; H_2O is astonishingly versatile as a solvent and participant in biochemistry and so it seems likely that extraterrestrial life would also be based on this compound. Perhaps, then, aliens may be drawn to the Earth for our wonderfully wet oceans and seas and rivers and lakes – to siphon off our hydrological cycle. The problem with this supposition is that there are loads of far better sources of water in space. In fact, we think that when Earth first formed from the swirling disc of gas and dust around the proto-Sun it was actually a pretty dry planet; the water to fill our oceans was delivered later by a barrage of comets and asteroids from the colder, outer regions of the solar system. In fact, Europa, one of the moons orbiting Jupiter, contains more liquid water in the global ocean beneath its frozen surface than our entire planet – Europa, and not Earth, is the Waterworld of our solar system. So if you were an alien voyaging between star systems in need of a drink, you'd have access to a far greater amount of water in the icy moons and cometary halo of the outer solar system. You'd also find it much more practical to operate in deep space, rather than trying to suck up the oceans against the gravitational pull of the planet Earth.

Aliens come to Earth for some other raw material. If not water, then maybe there's some other natural resource that aliens might invade the Earth to exploit. Perhaps they intend to wipe away our cities and begin strip-mining the crust of the planet for ores to extract metals and build more vast spaceships. But in fact, because the Earth formed from a molten state with iron sinking down to the core, our planet's crust is actually pretty depleted of useful metals like iron, nickel, platinum, tungsten and gold. And as with the water, it's hard to see why aliens would bother extracting material against the gravity of the Earth when the asteroids are composed of the same basic rocky stuff. In fact, some asteroids are believed to be essentially pure lumps of metal – they were once the cores of protoplanets that were smashed apart again by the colossal collisions in the early history of the solar system. Several companies are already proposing to launch asteroid mining operations to exploit these exceedingly valuable resources. Perhaps, though, there might be a reason that our hypothetical aliens would come to mine the Earth. While it's true that the asteroids and Earth, and other terrestrial planets, are made up of essentially the same rocky material, the Earth isn't simply an inert lump; it's a very active, dynamic place. In particular, the thin crust of the Earth is fractured into separate shards that are continually sliding around on top of the hot gooey mantle, rubbing alongside each other, crunching head-on, subducting one beneath another, or pulling apart to create fresh crust. This is the churning process of plate tectonics. So far, astronomers have already found over four and a half thousand extrasolar planets – worlds orbiting other stars – and the expectation now is that there are billions of rocky planets in our Galaxy. But here's a thought right on the forefront of current planetary science and astrobiology. Perhaps terrestrial planets are common, but terrestrial planets *with plate tectonics* are rare. Plate tectonics is thought to be vital for keeping the Earth's climate stable over billions of years to allow complex life like ours to evolve, and it also acts to concentrate certain metals into rich ores. It seems likely that only a small proportion of terrestrial planets undergo plate tectonics (neither Mars nor Venus does). So perhaps an alien civilisation would come to the Earth for our exceptional plate tectonics and concentration of particular metals, and the fact that the same tectonics had also enabled a rich biosphere to develop would be merely an inconvenience.

Aliens come to Earth looking for a new home. There is a considerable amount of rocky real estate in the Galaxy for aliens to consider moving home to, but as we shall see throughout this book, a terrestrial planet might need to offer more than just a habitable zone locale to be able to support complex life. Communities of hardy microbial cells thriving off inorganic energy deep underground might be able to survive pretty much anywhere, but complex life requires much narrower environmental conditions on the surface. Various features of the Earth beyond our warm oceans are thought to be crucial to maintaining a stable surface environment for geological time periods. These include plate tectonics acting to regulate the climate, a large moon preventing the spin axis of the planet from wobbling too much, and a global magnetic field for deflecting aside the solar wind and preventing the atmosphere being blown away into space. For these reasons, maybe planets like the Earth are something of a rarity, and so present particularly desirable targets for alien colonisation. But while it's true that such worlds may well be required for complex life to evolve in the first place, once an intelligent species becomes technologically advanced enough to travel between the stars it's also likely to be able to artificially manage a planet's environment. For example, many people are already starting to talk seriously about 'mega-engineering' or 'geoengineering' projects to avoid the worst effects of global warming on Earth, and we've worked out, at least in broad terms so far, how further in the future we could 'terraform' Mars to create a habitable environment for humans to live on the surface without needing spacesuits. Indeed, the very fact that Earth is already teeming with its own life (most of which is tenacious microbes that affect the chemistry of the atmosphere and oceans) may well be a hindrance to an alien species, with its own quirky biochemistry, looking for somewhere to colonise. It may well be easier to find a terrestrial world that hasn't already developed life of its own, and install its own biosphere on an empty planet.

Aliens come to the Earth for the Earthlings. To my mind, then, the enormous amounts of time and energy that are likely to be necessary for travelling between the stars in a Galaxy, and the fact that raw materials can be sought elsewhere more practically, would rule out aliens coming to the Earth simply to take something we have. I think we can safely rest assured that even if intelligent alien species do exist in our Galaxy, they are not about to appear in our skies with an invasion fleet to subjugate humanity and begin stripping our world. Perhaps the thing that may attract only extra-terrestrials to Earth is us. I suspect that if aliens did come to Earth, it would be as researchers: biologists, anthropologists, linguists, keen to understand the peculiar workings of life on Earth, to meet humanity and learn of our art, music, culture, languages, philosophies and religions.

If aliens do come to pay us a visit, there's one final way that the movies have probably got it all wrong. The laws of physics (at least as we currently understand them– after all, in a hundred years we may have worked out how to build a practical warp drive or stretch stable wormholes through the fabric of space-time) strongly constrain movement across the vast gulfs between stars. To make the journey time from one star system to the next anything less than scores of millennia, you need to accelerate your spaceship to a fair fraction of the speed of light. The greater the mass you need to accelerate, the greater the energy required, so you really want to keep your starships as small and light as possible.

Intelligent life forms like humans are inherently bulky things, particularly when you want to send a team of them along with all the life support machinery and regeneration systems for keeping them alive in space. But as Martin Rees has hinted in Chapter 1, a much more plausible alternative presents itself. Perhaps it's unrealistic to expect ET to go through all the discomfort and bother of actually voyaging in person across the oceans of interstellar space to far-flung worlds, but instead to travel by proxy. To cross the Galaxy not by encasing wet, vulnerable biological organisms *within* complex life-support technology, but *as* the hardened, durable technology itself. With a more complete understanding of how the human brain works – the neuronal wiring diagram and other interactions that give rise to intelligence and consciousness – it stands to reason that we could not only simulate this perfectly within hardware to construct an AI (artificial intelligence), but also potentially upload the consciousness of a living person into a computer.

Contained within a capsule of miniaturised electronics and systems for self-repair you'd not only be essentially immortal, but also incredibly compact and light and much better suited for interstellar travel. In this sense, perhaps most life in the Galaxy isn't carbon-based (organic), but silicon-based. I don't mean this in the sense of silicoid monsters imagined living inside volcanoes in *The X-Files* or *Star Trek*, but as the hardware supporting complex sentient computer programs. Silicon life would be second generation, existing only because it has been designed and created by a precursor organic species, which itself evolved naturally on a habitable world.

For these reasons, it strikes me that if there is intelligent alien life out there in our Galaxy, they almost certainly wouldn't pay us a visit in person in huge city-sized motherships, but by sending their sentient robots as emissaries. But how would they know we're here in the first place? Humanity has been leaking (or deliberately transmitting) radio waves out into space for roughly a century, so an alien civilisation running a SETI programme with sensitive radio telescopes could detect us. But this radio bubble announcing our technological emergence, centred on the Earth and expanding out into space at light speed, is only around 200 light years across. That is a minuscule region of space in the Galaxy as a whole, a disc 100,000 light years in diameter, and so even if the Galaxy does contain other intelligent life forms, they would likely still be oblivious to our recent appearance. But although humanity has only been detectably civilised for a century, the Earth itself has been conspicuously alive for many hundreds of millions of years, and this links to one of the hottest topics in current astrobiology.

Life on Earth, and specifically photosynthetic life such as plants and cyanobacteria that grow by absorbing the energy of sunlight and splitting water, has been releasing oxygen as a waste gas at such a high rate that it has built up in the atmosphere, first to just a few per cent, and today constituting a fifth of the Earth's air. Oxygen is a very reactive gas, and the only reason it has been able to accumulate in the atmosphere is that it is constantly being replenished by living organisms. In fact, the presence of oxygen in the atmosphere is thought to be so unusual to the geochemistry of a planet that astrobiologists consider it to be a biosignature of life (specifically if oxygen and a reduced gas like methane are both present). We are currently on the verge of building space-based telescopes that use spectroscopy to read the composition of the atmospheres of terrestrial exoplanets, and so survey the night sky for signs of life. And we're only relative newcomers on the galactic scene. There's nothing special about this exact moment in galactic history, and life on another planet could have evolved intelligence millions of years ago and used their own telescopes to look out for planets displaying the telltale sign of an oxygen-rich atmosphere. But apparently, as far as we can tell (and do have a read of Chris French's chapter on the psychology of UFO sightings) the fact that the Earth is obviously sporting biology has not prompted anyone to say hello.

This is a very curious observation, and to my mind could be down to two equally intriguing possibilities. The fact that Earth's oxygen-rich atmosphere has apparently attracted no one's attention may simply be because life is so rare that there is not a single other civilisation in the Galaxy with us to have their attention drawn. Or perhaps planets with an oxygen-rich atmosphere are so staggeringly common that the Earth just doesn't stand out among the masses. In the first possibility we are solitary and lonely intelligent beings in the Galaxy; in the second, life is absolutely rife in the cosmos. Both, to me, are equally profound realisations. And the most exciting aspect is that within your and my lifetime we will have launched our atmosphere-reading space telescopes and the science of astrobiology will have been able to tell which one is right.

What a time to be an intelligent life form on Earth!

3

Flying Saucers: A Brief History of Sightings and Conspiracies

Dallas Campbell

You're out walking the dog. It's late afternoon and getting dark. In the sky you see a bright light. Is it moving? You think so. You try and think what it might be, starting with the most likely: Aircraft landing light? Venus? Foil balloon reflecting the light? Iridium satellite flare? Could it simply be floaters in your eye? You are close to the nearby military base. Perhaps it's an exotic aircraft or one of those drones we hear so much about these days. Suddenly it dawns on you. Of course! It must be an alien scout craft from the Zeta Reticuli star system, piloted by three telepathic Greys with the tacit approval of a clandestine 'Majestic 12' US government group. It can only be a matter of moments before you are abducted, after which you will experience a feeling of paralysis, suffocation, missing time, and pain around the genitals. You will retain no memory of the event except during regression hypnosis, after which you will discover a small metal implant on the nape of your neck. It's bound to be one of them. Your dog barks excitedly at you in agreement.

I'm being a bit facetious. But only a bit. Such is the power of the human brain, and daunting is the task of the UFO investigator who attempts to untangle the Gordian knot of facts, fictions, fallacies and fallibilities that make up the thousands of UFO reports from around the world. Whether you're a believer, a sceptic or a 'want-to believer', it's easy (and fun) to get drawn into UFO culture, partly because the central idea of the neighbours popping round sounds quite reasonable. After all, Earth-based human civilisations meeting for the first time is a recurring and powerful theme in history, and we know that space is vast enough to plausibly contain other intelligent beings. Of course, pedants point out that the 'U' in UFO simply means 'unidentified' rather than alien, but for the rest of us the term UFO is inextricably linked with the *extraterrestrial hypothesis* (ETH): that unidentified flying objects are best explained by the presence of aliens. Like other supernatural explanations, you don't need to *believe* to feel a creeping unease in the telling of a ghost story on a dark winter's night in an old and creaky house.

In his book *The Flying Saucers Are Real*, the American journalist Donald Keyhoe investigated the late 1940s obsession with flying saucers and firmly concluded the Earth is indeed being observed and visited by alien civilisations. However, ufology – the study of such phenomena – is the very definition of 'not an exact science'. It's that perpetual elusiveness combined with a faint whiff of plausibility, mixed with a dash of paranoia, that fuels such ideas and makes them so widely appealing. Official denial by governments or incredulous questioning by sceptics is simply further evidence of a cover up and of a public blind to the truth. As the writer Jonathan Swift remarked: 'Reasoning will never make a man correct an ill opinion, which by reasoning he never acquired.'[1]

While UFOs might not yet actually have arrived on Earth, in our imaginations they have been everywhere, from the White House lawn in the movie *Mars Attacks* to the familiar, Roswell-inspired 'Grey' extraterrestrial, with its enormous almond-shaped eyes – so familiar to us these days that it has its own emoji. 👽 So, what were the events that breathed life into these images?

---

1 'Letter to a Young Clergyman' (1720).

Let us remind ourselves of five of the most notorious UFO stories that have taken the flying saucer from fringe subculture to mainstream modern folklore. I offer no possible explanations or comment on their validity, but merely present them, roughly, as reported. After all, the truth might be out there – but do we really want to find it? It's the mystery itself we enjoy. So wherever you are on the belief spectrum, forget it all for a few minutes. You can return to the science in a moment. Here's your handy cut-out-and-keep guide to the real flying saucer stories that changed the world.

The Kenneth Arnold Sighting

'Supersonic Flying Saucers Sighted by Idaho Pilot.'[2]

Picture a flying saucer in your mind. The mental image you now have was born on a beautifully crystal-clear Tuesday afternoon one minute before 3 p.m. on 24 June 1947. Idaho businessman and amateur pilot, Kenneth Arnold, was flying a CallAir light aircraft from Chehalis, Washington State, to Yakima over the Mt Rainer National Park, a distance of a little over 100 miles. En route he detoured, to see if he could spot the wreckage of a downed C-46 marine transport plane lost on Mt Rainer, for which there was a $5000 reward. Arnold couldn't find any wreckage, but moments later saw something that was to define the rest of his life: a series of bright mirror reflection-like flashes on his aircraft. The only other aircraft he could see at the time was a DC-4 some 15 miles away to his rear. Arnold then saw the cause: a 'chain of nine peculiar flying aircraft'. After ruling out geese he assumed they must be jet aircraft of some kind, but became increasingly frustrated at not being able to identify them. In his report to the Army Air Force Intelligence he described them as he'd observed – geese flying in 'a chain-like line'. He calculated their size using the DC-4 and a

---

2 *Chicago Sun*, 26 June. Source: Wikipedia.

tool he had in his pocket for reference, and their speed by timing the distance they covered between Mt Rainer and Mt Adams: over 1200 mph. Unheard of at the time.

Arnold landed at Yakima, where he told a friend about his strange sightings, before setting off for Pendleton, where he discussed what he'd seen with other pilots who offered suggestions of possible causes. Guided missiles? Experimental aircraft?

An excited media quickly got hold of the story. On 26 June the *East Oregonian* quoted Arnold's description of the objects in a number of different ways: 'flat like a pie pan' and 'somewhat bat shaped' and 'like the tail of a Chinese kite' and, most famously, 'flying disks' and 'saucer-like'.[3] The origins of the term 'flying saucer' itself became a source of contention which Arnold tried to clear up in a radio interview with Edward R. Murrow three years later:[4]

> These objects more or less fluttered like they were, oh, I'd say, boats on very rough water or very rough air of some type, and when I described how they flew, I said that they flew like they take a saucer and throw it across the water. Most of the newspapers misunderstood and misquoted that too. They said that I said that they were saucer-like; I said that they flew in a saucer-like fashion.

Whether misquoted or not, the popular image of the flying saucer was born. What's interesting is that Arnold makes no mention of the ETH in initial interviews. But with Murrow he says:

> I more or less have reserved an opinion as to what I think. Naturally, being a natural-born American, if it's not made by our science or our Army Air Forces, I am inclined to believe it's of an extraterrestrial origin.

---

3 *East Oregonian* interview with Kenneth Arnold, 26 June 1947. Source: Project1947.com.

4 Source: http://www.theufochronicles.com/2013/04/edward-r-murrow-interviews-kenneth.html and http://www.project1947.com/fig/kamurrow.htm.

The Roswell Incident

'RAAF Captures Flying Saucer on Ranch in Roswell Region'[5]

This was the July 1947 headline that sparked a multi-million dollar alien industry. For an in-depth analysis of what happened that day and how the story evolved over the years, you must carefully pick through endless articles, books, TV documentaries, and approximately 50 per cent of the entire internet which seems devoted to the subject. In brief: rancher William 'Mac' Brazel found strange debris strewn over a field situated to the north-west of Roswell, a small city in New Mexico. Thinking it might be wreckage from the recently publicised 'flying disks' (the infamous Kenneth Arnold sighting was only a few weeks before), he reported it to the Sheriff, who reported it to the military – the 509th Bomb Group at Roswell Army Airfield – and the recovery of this debris was overseen by Major Jesse A. Marcel. The 'Captured Flying Saucer' news story was released to much excitement – and then later retracted after the mysterious debris had been moved to Fort Worth Army Air Field. Marcel was famously photographed squatting down, holding up pieces of the debris (now officially reconfirmed as some kind of radar reflector and weather balloon array). That at least seems to be roughly agreed upon. But was the material switched? Was this story retraction a sign of a clandestine government cover-up? Was this material shed from a crashed saucer(s) in another location? In the late 1970s, the dying embers of the story were rekindled, igniting into a full-blown conspiracy inferno by the 1990s involving various ufologists, most notably Stanton Friedman (*Crash at Corona, Top Secret/Majic*).

It's thanks to tenacious investigators like Friedman that you know what Roswell means: from some foil, balsa wood, sticks and scotch tape in a field, and an overenthusiastic subeditor, we've

---

5 *Roswell Daily Record*, 9 July 1947. Source: Wikipedia.

arrived at exotic other-worldly materials with strange properties and alien hieroglyphs, a disputed number of crashed saucers, government cover-ups, recovered alien bodies, alien autopsies caught on film, documents from the clandestine committee named Majestic-12,[6] and a film starring Ant and Dec. The US Air Force published *Roswell Report: Case Closed* in 1997, citing the spy-balloon programme *Project Mogul* as an explanation.

Today, although drifting into parody, Roswell is an important part of modern American folklore and a potent symbol of government mistrust. In 2016, Hillary Clinton even made it an election pledge to investigate the government's involvement in UFOs,[7] such is the huge and presumably vote-winning fascination. Both Bill Clinton and Barack Obama appeared on the *Jimmy Kimmel Live!* chat show joking about their involvement in the subject, Clinton confirming he had ordered a review of the Roswell documents during his second term. Kimmel asked Clinton about Roswell and Area 51 (next up), 'If you saw that there were aliens there, would you tell us?' 'Yeah ... I would,' replied Bill to rapturous applause. But then he would say that, wouldn't he.[8] 'Case closed' is something that Roswell can never be.

---

6 'Majestic 12' is the alleged name favoured by conspiracy theorists for a government group of politicians and scientists who have knowledge of an alien agenda. Stanton Friedman's *Top Secret/Majic* is the book in which he claims to expose their identities.

7 http://www.huffingtonpost.com/entry/
hillary-clinton-vows-to-investigate-ufos_us_5687073ce4b014efeoda95db

8 And here's Bill Clinton joking about it all: https://www.youtube.com/watch?v=gZqLIHRepSo.

Area 51

'Area 51 and extraterrestrial life both exist, says head of NASA.'[9]

… just not in the same place. Welcome to the world's most famous top-secret military base. Area 51 – a government storage locker for our most secret knick-knacks is more a state of mind than a physical location. In the last 25 years, along with Roswell, it has become synonymous with all things alien. Now firmly rooted in mainstream pop culture it has become a science fiction location for movies like *Independence Day* and countless TV programmes. It is even the official storage facility of the Ark of the Covenant.[10] For decades the US military wouldn't even acknowledge it, but these days you can enjoy whizzing over it via Google Earth. Situated on Groom Lake, a dry lake bed in the middle of a vast area of restricted government land in the Nevada desert, it is hidden away from prying eyes behind a line of hills in the charmingly named Tikaboo valley. Despite its new fairy-tale status it is still a heavily guarded facility. Built as a remote Air Force location to test the Cold War generation of classified 'black' stealth aircraft like the SR-71, U2 and F117, the areas surrounding it developed a second life in the 1980s as a popular 'saucer-watching' spot for reasons you can probably guess. Out here on one of America's loneliest roads, the 375 'Extraterrestrial Highway', stories, sightings, and conspiracies breed and cross-pollinate unchecked.

There's an interesting cast of characters who put Area 51 on the map in the early 1990s, in particular Bob Lazar, who was interviewed by Las Vegas television reporter George Knapp in 1989 under the pseudonym of 'Dennis', and again later under his real

---

9 Sarah Knapton, *Daily Telegraph*, 19 June 2015.
10 See the disappointing *Indiana Jones and the Kingdom of the Crystal Skull.*

identity.[11] Lazar claimed to have worked for a few months as a 'senior staff physicist' at 'S-4', an even-more-secret secret annex of the Area 51 Groom Lake base (complete with Bond-style camouflaged hangar doors built into the hillside) at nearby Papoose dry lake. Under above-top-secret 'Majestic' clearance he worked on a project reverse-engineering flying saucers to understand how they worked, with a particular speciality in propulsion systems using 'element 115' in some kind of antimatter drive. Lazar describes the nine saucers he saw as being like brushed aluminium without seams or welds, as if injection moulded. He also witnessed a test of one of the saucers over the lake bed. Although he never saw aliens, Lazar apparently saw alien autopsy photographs in briefing documents.

Lazar's employment and education background have been widely discredited; no record of the prestigious degrees he claimed to hold can be found. This, along with various physical threats, is, he argues, clear evidence of the authorities trying to erase his identity.

Rendlesham Forest Incident

'UFO lands in Suffolk. And that's official!'[12]

Dubbed 'Britain's Roswell', this incident took place over two nights at the twin Suffolk RAF bases Woodbridge and Bentwater, used by the US Air Force at the time. It's got all the classic ingredients: a military context, multiple credible eyewitnesses, 'Men in Black' cover-ups, sworn testimonies, interrogations with big syringes. But was it an extraterrestrial encounter over a nuclear

---

11 George Knapp on Bob Lazar: https://www.youtube.com/watch?v=1i4Fu_L7cHM.
12 *News of the World*, 2 October 1983. Source: http://www.ianridpath.com/ufo/headline.htm.

facility? Or just the nearby Orfordness lighthouse? Reports, state-
ments and interviews have shifted over time, but the basics of
the story go something like this: On 26 December 1980, in the
wee hours, an American Air Force security patrol led by Jim
Penniston went to investigate mysterious lights in Rendlesham
Forest, thought to be a crashed aircraft. But Penniston noticed
something odd: there was no smell of burning fuel or wreckage,
and their radios became scrambled. In front of them they saw a
bright white light moving among the trees along with smaller blue
and blinking red lights. In later interviews Penniston describes a
triangular mechanical object, with sides of about 9 ft and 7 ft,
completely silent, with no sign of engines or crew compartment.
He reports touching it: 'black, smooth and glass-like'. On one
side he notices markings, not any known language but shapes or
glyphs (perhaps reminiscent of the Roswell debris markings). At
2:45 a.m. the object begins to hover a few feet in the air, still silent,
then vanishes upwards. Landing indentations were later identified
on the ground.

Two nights later Deputy Base Commander Col. Charles Halt
leaves the base with other personnel to investigate a possible return
of the object. He carries a Dictaphone, narrating events as he sees
them, as well as cameras and a Geiger counter. The UFO appears
to return: a red light is seen weaving through the trees. The light
moves into a farmer's field. Witnesses describe seeing what looks
like molten metal shedding off the object. The light breaks into
smaller white objects which then disappear at great speed.

Halt writes up his story in 'The Halt Memo', an official
Ministry of Defence letter, which he later retracts as a sanitised
version of events. Other witnesses file their reports; stories are
amended and embellished as the years roll by. Security policeman
Larry Warren claims to have witnessed the second sighting from
another vantage point, interpreting the strange lights as a craft of
some sort and even seeing 'beings'. Witnesses claim to be inter-
rogated and persuaded to sign official documents by mysterious

dark-suited figures from the Office of Special Investigations who some consider to be the fabled 'Men in Black' – government agents whose job it is to silence UFO witnesses. Penniston claims, through regression hypnosis, to have been injected with the 'truth serum' sodium pentothal by them.

The lighthouse remains the most favoured explanation, but this bottomless story continues to appeal – not least because many of the witnesses (unlike Roswell) are still alive, and 'Fortean' writers such as Ian Ridpath and Jenny Randles are still commenting on the story.[13]

The Betty and Barney Hill Abduction Case

'A UFO chiller. Did they seize the couple?'[14]

It's the last thing you want to happen to you when you're driving late at night on a rural road in New Hampshire: Betty and Barney Hill and their dog Delsey were heading home from a vacation in Canada. Betty was a social worker, and Barney worked for the post office and was involved in a local civil rights organisation. It was about 10:30 at night on 19 September 1961. Betty saw a light in the sky which seemed to be following them, moving erratically. A new planet? A shooting star? An aircraft? Slowing down to get a better look, it seemed to be a disc-shaped craft with flashing lights. The Hills eventually pulled up in the middle of the road with the pancake-shaped spinning craft hovering in mid-air 100 feet in front of them, and about 50 feet in the sky. Barney got out

---

13 Much of this testimony comes from witness interviews from various television documentaries. For more detail on the story and analysis, I recommend www.ianridpath.com who has a huge amount of sensible information about it.

14 *Boston Traveller*, 25 October 1965. Source: Wikipedia.

of the car and claimed to see through his binoculars a group of small humanoids in black uniforms watching him from windows around the perimeter of the craft. Terrified, he ran back to the car and they drove off. As they were driving they heard a series of strange beeping sounds from the back of the car. The Hills then reported entering a drowsy altered state, finally arriving home some two hours later than expected at 5:15 a.m. Parts of the journey and chunks of time were missing or blurred in their memories. They reported memories of leaving the highway onto a dirt road and there being a road block, human figures and a glowing orb of some sort. At home Betty noticed a strange pinkish powder on her torn dress. Barney's shoes seemed mysteriously scuffed. He felt an irritation around his groin. There were strange polished spots on the boot of the car, which they discovered would send a compass needle spinning.

A few days after the incident Betty experienced a number of vivid dreams that pieced together a narrative of events: Barney had driven off the main road into a wooded area where they encountered a group of small humanoid beings who led them onto the landed craft. Betty described them as being around 5 feet tall with large eyes, a thin slit for a mouth, and no protruding ears. They spoke a limited form of broken English. On board, and after some protest, Barney and Betty were taken away for separate medical examinations – tests apparently to distinguish the differences between the alien humanoids and human beings. She described being taken into a bare room where her ears, nose, throat and eyes were examined, and hair, nail and skin samples were taken. A large needle was inserted into her navel, causing great pain, which they promptly stopped. Betty engaged in conversation with one of the beings, who at one point showed her a detailed galactic star map.

Two years later the Hills had their first medical hypnosis sessions with Dr Benjamin Simon, a psychiatrist specialising in military psychological trauma. Barney had been referred to him by another doctor to try and help with ulcers, anxiety and stress. Barney's sessions uncovered similar themes to those in Betty's dreams, including his own – more intimate – medical examinations and a deeply unnerving sense of telepathy between himself and his abductors. Betty's star map also came to light in these sessions and was subsequently reconstructed: the dots were joined and concluded by some to be that of the Zeta Reticuli binary star system. The whole story is recorded in John G. Fuller's book *Interrupted Journey*, which brought the phenomenon of *alien abduction* to world wide attention.[15]

These, then, are five of the UFO stories that launched a million sightings. Thank you for your attention. You're now free to go. Please turn the page and continue on your journey. In the words of my much missed mentor and theatrical 'Fortean' seeker Ken Campbell: 'I'm not mad, I've just read different books.'

_____

15 Letters, notes and articles from this case, involving Ben Simon and skeptical UFO researcher Philip Klass, can be viewed on Robert Sheaffer's website: http://www.debunker.com/historical/BettyHillBenjaminSimonPhilipKlass.pdf.

4

Aliens on Earth: What Octopus Minds Can Tell Us about Alien Consciousness

Anil Seth

You don't need to go to outer space to encounter an alien. To find other-worldliness here on Earth, go meet an octopus. A few years ago I spent a week with a dozen octopuses at a marine biology station in Naples, a guest of the biologist Graziano Fiorito. Like many others fortunate enough to spend time with these remarkable creatures, I was left with a vivid sense of an intelligent presence very different from our own.

When we think about aliens we usually think about strange body shapes, unusual abilities and uncanny intelligence: whatever kind of intelligent life is out there, it's likely that alien consciousness would be very different from our own. So say hello to the extraordinary octopus, our very own terrestrial alien. Eight prehensile arms lined with suckers; three hearts, an ink-based defence mechanism, and highly developed jet propulsion; a body that can change size, shape, texture and colour at will, and cognitive abilities to rival many mammals. The common octopus, *Octopus vulgaris*, has about half a billion neurons, roughly six times more than a mouse. Strikingly, the majority of these nerve cells are found, not in its central brain, but in its semi-autonomous arms which are almost like independent animals. There are many signs of octopus intelligence: they can retrieve hidden objects – usually tasty crabs – from within nested Plexiglas cubes, find their way through complex mazes, utilise natural objects as tools, and even solve problems just by watching other octopuses do the same. Even their DNA seems out of this world: 'It's the first sequenced genome from something like an alien,' said neurobiologist Clifton Ragdale in the journal *Nature*. So, if there are sentient aliens out there, somewhere in the Universe, one way of trying to understand what sort of consciousness they may have is to think about the inner Universe of the common octopus.

Defining consciousness

To do this, we need a working definition of consciousness – and herein lies the first problem, since there is no well-established definition on which everyone agrees. A simple approach is to say that for a conscious organism *there is something it is like to be that organism*. Or one can say that consciousness (at least for us) is what disappears when we fall into a dreamless sleep, and what returns the next morning when we wake up. Putting things a bit more carefully, for conscious organisms there exists a continuous (though interruptible) stream of conscious scenes or experiences – a 'phenomenal world' – which is subjective and private.

Taking humans as a benchmark, we can draw some further distinctions. The first is between conscious *level* and conscious *content*. Conscious level refers to how 'conscious' an organism is. This can be thought of as a graded scale from complete unconsciousness (such as happens under general anaesthetic), all the way to vivid conscious wakefulness. Importantly, conscious level is not the same as wakefulness: one can be conscious while asleep, for example while dreaming, and one can be unconscious while physiologically awake, as in sleepwalking and in some medical conditions like the vegetative state.

Conscious *contents* refer to the elements of a conscious scene:

they are what you are conscious *of*, when you are conscious. Conscious *content* includes (again, for humans) colours, shapes, smells, thoughts, explicit beliefs, emotions and moods, experiences of desire and agency, and so on. Collectively, conscious contents are what philosophers call 'qualia' and explaining how qualia emerge from physical 'stuff' remains the most metaphysically mystifying aspect of studying consciousness. Conscious contents can be divided again into those that are *world-related*, like the smell of freshly cut grass, and those that are *self-related*, like the ache of a rotten tooth or the experience of identifying with a particular body. Some aspects of self-consciousness, like experiences of body ownership and of having a first-person perspective on the world, are so continuous and pervasive that it is easy to take them for granted. But it's just these aspects of consciousness that might be most different in species like *O. vulgaris*, since their bodies and the way they interact with the world are so very different from us.

Conscious level: Are octopuses conscious at all?

The most basic question is whether an octopus is conscious at all. Let's approach this by looking at what is necessary for human consciousness, and then seeing whether octopuses share these essential mechanisms. In humans, consciousness is not simply a matter of having lots of neurons. The human brain contains about 90 billion neurons in total – an unimaginably vast number. Strikingly, the majority of these neurons are found in the cerebellum, the small 'mini brain' hanging off the back of the cortex. This part of the brain, while important for all sorts of things, does not seem to be necessary for consciousness. In fact, consciousness cannot be traced to any single region within the human brain. It's true that there are some regions that if damaged will abolish consciousness forever, but these are better understood as 'on/off switches' than actual 'generators' of conscious experiences.

Our current best guess is that human consciousness depends on how different brain regions talk to each other. When consciousness fades, for example under general anaesthesia or in deep dreamless sleep, the brain becomes functionally disconnected. Its various regions become increasingly isolated, leading to an overall loss of integration. The opposite can also lead to a loss of consciousness. During epileptic absence seizures, global brain activity becomes highly synchronised as electrical storms erupt across the cortex. Many experiments now show that during normal wakeful consciousness different brain regions can to some extent do their own thing, while at the same time participating in an integrated 'whole'. This makes sense from the point of view of what conscious experiences are like for the experiencer: every conscious scene is experienced as unified, and at the same time is composed of many different elements and is different from every other conscious experience. As one popular neuroscientific theory puts it: conscious experiences contain a great deal of 'integrated information'.

Could an octopus be conscious on this basis? The half a billion neurons within the *O. vulgaris* nervous system would seem easily enough to provide a large repertoire of possible conscious contents. But the octopus nervous system is much poorer in the fast, long-range connections that connect the diverse regions of the human brain. Octopuses lack myelin, the insulating material that enables these long-range connections to develop and function. And as already remarked, the majority of an octopus's neurons lie outside the central brain, a situation quite different from the nervous systems of mammals. What this suggests is that the integration of the activity of different brain regions, necessary for consciousness in humans and other mammals, might be very different in octopuses. This doesn't mean that octopuses necessarily lack consciousness; it just means that their consciousness might have a very different character to ours. It might be less integrated into a single well-defined scene, or there might even be multiple partial consciousnesses overlapping within a single body, as has been speculated to happen in people who have had their brain hemispheres surgically separated (so-called 'split brain' patients), a technique once used to treat severe epilepsy.

Definitive evidence for octopus consciousness is still hard to come by. At a behavioural level, octopuses, like most other creatures, go through cycles of waking and sleep, and they also respond to anaesthetics like isoflurane at similar dose concentrations to other species. But at the neural level we know very little. There have been very few direct brain recordings from octopuses, and those that have been done have mainly looked at single neurons involved in learning and memory. What's needed now is to record neural activity from large parts of the octopus brain during different states of physio-logical arousal (and also under anaesthesia), to see whether we can find those characteristic patterns of balanced differentiation and integration that we see with human consciousness.

Conscious content: What are octopuses conscious of?

If octopuses *are* conscious, what might they be conscious of? Let us again consider the human case first, and then make some comparisons with the octopus.

Human conscious contents are very varied, ranging from those associated with sensations of the outside world, to emotions, moods, beliefs and thoughts, experiences of 'will' and volition, and many others besides. Focusing on sensory awareness, the classic human senses are vision, hearing, touch, taste and smell. These are accompanied by less well-known but equally important sensory channels, including senses of body position and movement ('proprioception' and 'kinaesthesia'), balance, pain, temperature, and a whole raft of inputs reflecting the internal state of the body, such as hunger, thirst, cardiac activity, and the like.[1]

---

1 These senses, that reflect internal physiological state, are collectively

What about the octopus? In terms of sensory capabilities, all octopuses have good vision, even for the low light conditions prevalent at night or on the ocean floor. Amazingly, octopuses can also 'see' with their skin, helping them match their immediate environment for purposes of camouflage. Octopuses also share the classic senses of taste, smell and touch – and they can hear, but not very well. Octopus arms are particularly rich in sensory receptors, and not just for touch – the many suckers also provide a sense of taste. When the whole genome of *Octopus bimaculoides* (the Californian two-spot octopus) was recently sequenced, it revealed a set of octopus-specific genes that are expressed within the suckers and which are associated with a specific sort of neuro-transmitter (acetylcholine) that may be the source of this unusual ability. Of the less familiar senses we know very little, though it is highly likely that octopuses directly sense the state of their body in various ways. They certainly have pain receptors and show a range of pain-related behaviours similar to vertebrates, including grooming and protecting injured body parts.

Perception isn't just about having this or that sense. When we perceive our environment, for example using vision, we don't just build an accurate picture of an objective reality, like an internal camera might do. Instead, we perceive the world in terms of how we might act in it and on it. A door, for instance, is perceived as 'something that can be opened' and not just as a rectangular slab of wood. Since octopuses (and any aliens out there!) have very different sets of possible actions to you or me, this means that they will likely have very different perceptions, even if they are in the same environment and have the same senses.

------

called 'interoception'. The role of interoception in brain function and especially in consciousness has not received as much attention as perception of the outside world. This is now changing, with some researchers (including me) thinking that interoception may be more fundamental to consciousness, and especially to self-consciousness and subjectivity.

Even though direct evidence about perceptual consciousness in octopuses hasn't yet been found, the fact that octopuses
display impressively flexible behaviour does suggest that they
may indeed have conscious perception. In humans, consciousness is closely associated with behavioural flexibility (like when
we decide whether to avoid, or to eat, an unusual object), whereas
many purely instinctive reactions (like snatching your hand away
from a hot stove) don't require consciousness. In other words,
a conscious octopus wouldn't just respond to its environment
instinctively: octopuses process the information they receive and
make decisions about it.

Conscious self: What is it like to be an octopus?

A crucial feature of human consciousness is the variety and sophistication of our *self-consciousness*. Human self-consciousness – the
experience of being 'me' – plays out at many levels. These include
a basic sense of being and having a body, to experiences of looking
out onto the world from a particular first-person perspective, and
then to experiences of volition and 'will'. There are also higher-
level aspects of selfhood which have to do with the continuity
of self-experience across time: these include autobiographical
memories of specific events and the concept of 'I' to which we
attach a particular name ('Anil', for me). Human selfhood is also
intrinsically social: the way I experience *being me* depends to some
extent on how I think you perceive me.

Let's focus on just one of these levels of self: the experience of
identifying with a particular body. It may be tempting to take this
for granted, but it would be a mistake to do so. Some neurological
conditions involve severe disturbances of embodied selfhood. For
example, *somatoparaphrenia* involves the experience that one of
your limbs belongs to someone else, and many amputees still feel
pain in their missing ('phantom') limb. Altered body experience

can also be induced in much less dramatic situations. In the famous 'rubber hand illusion', a person's real hand is hidden from view while they fix their gaze on a fake rubber hand. If both hands (real and rubber) are then simultaneously stroked with a soft paintbrush, most people have the bizarre experience that the rubber hand somehow becomes part of their body. This shows that our experience of what is, and what is not, part of our body is not simply given, but is a surprisingly flexible perception generated by our brain.

If an octopus experiences its environment differently from us, its experience of its own body is likely to be even more bizarre. To begin with there is its strikingly decentralised nervous system. Delegating some 'neural' control locally to each arm makes sense because octopus arms can move in many different ways: they are far more flexible than our relatively stiff jointed limbs. Many studies have indeed shown that octopus arms are capable of behaving semi-independently and can execute complex grasping movements even after being severed from the body. This suggests that in general the octopus might have only a hazy experience of its body configuration, and that there might even be something it is like to be an octopus arm!

Tentacles are often a feature of aliens as imagined in science fiction, and the flexibility of octopus arms poses a distinctive challenge that might apply to strangely shaped aliens as well: how to avoid getting all tangled up. For octopuses, the suckers on each arm automatically grip onto almost any passing object, but somehow they do not fix onto any of the other arms (or the central body) despite almost continuous contact. One way to achieve this feat of self-discrimination would be for the central brain to maintain an up-to-date picture of the position of each limb. This is a hard enough problem for humans, but our brains seem to be up to the job. For an octopus the problem seems formidable. Instead, it has recently been discovered that octopuses secrete a chemical in their skin, which prevents the suckers of other arms from attaching, and which constitutes a highly distinctive chemical-based self-recognition system. This just shows that even here on Earth, being a conscious self may involve types of sensation that are completely alien to us humans – making the task of imagining conscious aliens even more challenging!

The novelty of octopus embodiment doesn't stop with semi-autonomous arms and chemical self-recognition. Octopus bodies can undergo dramatic and rapid changes in size, shape, colour, patterning and texture. They have remarkable camouflage abilities, blending seamlessly into the environment while waiting for tasty prey to swim by (or predators to move along). Putting all this together means the experience of body ownership for an octopus might well be the most other-worldly aspect of its consciousness.

Aliens on Earth and beyond

Is consciousness a unique event in the history of the Universe, concentrated by evolutionary happenstance on a small planet in a distant backwater of a remote Galaxy? Or is consciousness here, there and everywhere? Perhaps consciousness is even a fundamental property of the Universe itself, like electrical charge and mass. As yet, nobody knows.

What we do know is that we humans are conscious, and that we share many of the biophysical mechanisms needed for consciousness with other animals including non-human primates, other mammals, and perhaps even non-mammalian species like birds and octopuses. As with many complex biological features, consciousness also seems to serve a useful purpose. For humans, this may be by providing us with a very large amount of integrated information, organised within a conscious scene, so that we can 'do the right thing at the right time' in a complex and ever-changing environment.

If this evolutionary story is on track, it follows that consciousness is likely to be a property of evolved life forms wherever in the Universe they have arisen, once certain thresholds for complex behaviour have been crossed. Importantly, these thresholds might not be very high. Being conscious is not about being able to engage in rational thought or to use language, it is fundamentally about perceiving the world – and the self – in a way that ensures the survival of the organism in a world full of danger and opportunity. On this view, consciousness is more likely to be a property of even simple lifeforms with nervous systems engaged in flexible self-preservation than it is to emerge from complex robots or artificial intelligences that can mimic advanced human abilities – like playing 'Go' – but which fundamentally do not care about their own existence. The experience of *being embodied* might therefore be one of the most basic of all conscious experiences, and the one most likely shared by aliens, terrestrial or otherwise.

Many years ago the philosopher Thomas Nagel famously asked 'What is it like to be bat?', highlighting the so-called 'explanatory gap' between the objective descriptions of science and the subjective 'what-it-is-like'-ness of consciousness. He was right to point out that scientific descriptions alone can never enable us to experience an alternative consciousness. We humans are forever trapped within the inner Universes prescribed by our own brains, bodies and environments. But by studying the limits of our own awareness alongside the remarkable abilities of other species, and by realising that the way *we* experience the world, and the self, is not the only way, we can gain startling glimpses into a space of 'possible consciousnesses'. We might never experience what it is like to be an octopus, but it seems very likely that there is *something* it is like to be this terrestrial alien.

As for extraterrestrials, wherever they may be, the possibilities are even more tantalising. Think of the strange environments on far-flung planets that would require totally new kinds of sensation. Consider the bizarre body types, perhaps even made of materials like silicon, that these environments might support. At the furthest reaches of our imagination we may find disembodied intelligences or 'hive minds' supporting a consciousness that is spread out across multiple individuals, so that there is no single 'I'. What is certain is that the inner Universes of consciousness – whether for you, me, an octopus or an alien – are every bit as fascinating and mysterious as anything we might find out there among the stars.

5

Abducted: The Psychology of Close Encounters with Extraterrestrials

Chris French

Many of the contributions in this book, each written by an eminent expert in his or her own field, discuss the possibility that life may have evolved elsewhere in the Universe. It is almost universally agreed that if contact with alien forms of life is ever established it would be one of the most sensational scientific discoveries that humankind could ever make. The implications for our view of ourselves and our place in the Universe would be profound. It is not surprising therefore that this is an issue that fascinates us and provokes intense speculation. However, there are also millions of people around the world who believe that such speculation is a complete waste of time. They believe that there is already convincing evidence to show that aliens not only exist, but have already made contact with humans.

A National Geographic Society survey of 1114 Americans in 2012 found that 36 per cent believed that 'UFOs' exist and only 17 per cent did not. The rest were undecided. A little caution is necessary here as the survey did not explicitly equate UFOs (unidentified flying objects) with aliens, but it seems reasonable to assume that most respondents would have implicitly equated the two. By extrapolation to the US population as a whole, this means there are some 80 million Americans who believe that most of the contributors to this book have wasted their time merely *speculating* about the chances of life 'out there'. Indeed, one tenth of the respondents in the survey claimed that they themselves had seen a UFO.

Similar results are found with UK respondents. According to a 2014 poll of 1500 British adults and 500 British children (aged 8 to 12), 51 per cent of the adults and 64 per cent of the children believe in aliens and 42 per cent of the adults and 50 per cent of the children believe in UFOs. It is of course perfectly sensible to make the distinction between belief in the existence of aliens and belief in UFOs. It is perfectly reasonable to accept that alien life may have evolved elsewhere in the cosmos – but is it also reasonable to accept that aliens are currently visiting Earth on a fairly regular basis? Or can the UFOs that some claim to have seen, and that others have heard about second-hand but nevertheless believe to be true, have more mundane – terrestrial – explanations?

J. Allen Hynek was an American astronomer who acted as scientific adviser to the US Air Force on a number of projects investigating UFO claims from the 1940s to the 1960s. Initially sceptical, he famously and controversially changed his mind on the subject, stating that there was 'sufficient evidence' to defend both the ET hypothesis (the idea that UFOs were extraterrestrial spaceships) and the even more controversial notion that UFO reports may be evidence of intelligence originating in 'other dimensions'. He went on to produce the classification system for types of UFO contact made famous by Steven Spielberg's 1977 blockbuster, *Close Encounters of the Third Kind*.

In this chapter, however, I will argue that psychological factors provide plausible explanations for close encounters of all kinds. When it comes to weighing the evidence for and against any particular claim, one of the most important psychological factors is that of *confirmation bias*. Confirmation bias is arguably the most pervasive cognitive bias to affect our thinking. It refers to the tendency that we all have to interpret the available evidence as supporting what we already believe or would like to be true. Given the undoubted appeal for many of us of the idea that we are not alone in the Universe, it should perhaps not surprise us that even poor-quality evidence is enough to convince many people that aliens have already made contact.

Close Encounters of the First Kind

The first type of close encounter in Hynek's classification system is simple sightings with no other supporting evidence. Ever since human beings first looked up into the sky, there have always been occasions where some unidentified object (or perhaps meteorological or celestial phenomenon, such as a bright meteor burning up in the atmosphere) was spotted. Of course, if the phrase 'unidentified flying object' were used simply to suggest that the viewer didn't know what they were looking at, it wouldn't be a problem. But to modern minds, the phrase is synonymous with the idea of an extraterrestrial craft of some sort.

In fact, even those who defend the ET hypothesis readily acknowledge that the vast majority of sightings have relatively mundane explanations. The most common causes of UFO reports include bright stars and planets, meteors, aircraft seen from unusual angles, laser displays, weather balloons, and Chinese lanterns. Usually, it's possible to work out what the UFO really is by checking the reported time of the sighting and its position in the sky against its probable cause – a night flight taking off from a nearby airport, for instance.

What is interesting in such cases is the way in which reports are often influenced by what the observer *thinks* they are seeing, in contrast to what they *are* actually seeing. Bear in mind that we usually make judgements about the size, distance and speed of objects by comparing them to the size of other objects that are nearby. If we see an unknown object in the sky, it is very likely that such cues will not be available. The image on the retina of a large object that is far away and moving fast is exactly the same as a smaller, closer object that is moving more slowly – and yet observers often confidently report the size, speed and distance of UFOs. No one is immune to such perceptual errors. There are well-documented cases of professional pilots reporting objects flying past their planes within a few hundred metres that further investigation revealed to be meteors literally hundreds of miles away.

Proponents of the ET hypothesis often adopt a position that seems to imply that if sceptics cannot definitively account for every single case that is presented to them, they should accept the ET hypothesis. This is simply unreasonable. Just as the police cannot solve every case they investigate, some UFO sightings will forever remain unsolved simply through lack of relevant evidence. The burden of proof in science always lies with the claimant, not with those doubting the claim.

Close Encounters of the Second Kind

The second of Hynek's categories refers to cases that involve some kind of physical evidence, usually photographic or video evidence but also including marks on the ground and/or increases in radio-activity at alleged landing sites and even radar recordings. I will limit myself here to commenting upon photographic and video evidence, but suffice it to say that other forms of physical evidence can also usually be plausibly accounted for without resorting to the ET hypothesis.

The statement 'The camera never lies' has never been true and, in the age of Photoshop, it has never been less true. Many classic photographs of alleged UFOs have been shown to be deliberate hoaxes, but it is probably the case that most claims to have caught a UFO on camera are by people who genuinely believe in their sighting. There are two other ways in which someone can come to sincerely believe that they have photographic evidence of an alien spaceship. The first is simply an obvious extrapolation from our discussion of close encounters of the first kind. Someone sees something up in the sky and manages to take one or more photographs of it. Of course it may later be correctly identified and its true, less exotic, nature revealed if appropriate further investigation is undertaken, but this may never happen, and the photographer is left convinced by their 'evidence' of the close encounter.

The second possibility is that the 'UFO' is not noticed at the time the photograph was taken, but only on later examination. Two interesting psychological phenomena are of relevance here. The first is known as *inattentional blindness* and refers to the fact that we often fail to notice something that is clearly present in our field of vision if we are concentrating on something else.

The classic demonstration of this effect involves showing a short video clip of two groups of people, one group in white shirts and the other in black. The members of each group are intermingled and, within each group, a ball is thrown to other group members. Part way through the clip, a person in a gorilla suit enters the scene, stands in the middle of the group beating her chest for several seconds, and then exits. Unsurprisingly, if viewers simply watch the scene without doing anything else, everyone sees the gorilla. However, if instructed to focus on counting the number of times the people in the white shirts throw the ball, ignoring the people in the black shirts, almost half of the viewers fail to see the gorilla at all. This well-documented and extremely reliable result completely undermines our feeling that something so bizarre should automatically catch our attention. For the same reason, if the UFO photographer is focusing their attention on the central subject of the photograph, they may well fail to notice something unusual in the background until the photograph is examined later. So something that shows up in a photograph as a mysterious silver blur could have been correctly identified and explained away at the time – had we not been distracted – as, say, a passing hot-air balloon.

The second psychological phenomenon of relevance here is

pareidolia. This refers to the tendency we all have to sometimes perceive random patterns as clear and distinct objects, such as seeing faces in clouds, in the grain of wood, or even on pieces of cheese on toast. In the context of photos of alleged UFOs, this means that some perfectly mundane object that was in the background and either moving very quickly or else photographed from an unusual angle may well end up producing an ambiguous image that might be mistaken for a flying saucer or some other type of alien craft.

It is always possible, of course, that some of the photographic or video evidence presented might really be a record of a genuine alien craft that has entered our atmosphere – or that such a visitation will be recorded in this way at some point in the future. True scepticism requires us to keep our minds open to such a possibility. But it is striking that, despite the ubiquity of CCTV security cameras and good quality cameras on mobile phones, the photographic evidence to support UFO claims is generally just the same as it ever was – mostly blurred and shaky images of unidentified lights against a background of dark sky.

Close Encounters of the Third Kind

The title of Spielberg's famous film refers to direct contact between humans and aliens. In 1952, George Adamski claimed to have met an attractive female alien in the Californian desert and to even have been taken for a ride in her spaceship. He was the first of the so-called *contactees* of that era, each of whom wrote many best-selling books about their adventures with the friendly aliens. Such accounts, although entertaining, were not even taken seriously by the ufologists of the day, who generally felt that they represented the 'lunatic fringe' and would bring ufology into disrepute.

For a while, there was a continuous stream of contactees ever ready to give media interviews (and sell books) regarding their encounters with enlightened aliens. In the early years, they often produced photographs of UFOs to support their claims, many of which were quickly dismissed as deliberate hoaxes. Perhaps in response to such debunking, within a few years it became more fashionable to claim that contact was taking place via a psychic connection rather than by visitations from physical spaceships. The contactee would typically apparently enter a trance state and be 'taken over' by the alien in order to deliver their messages to Earth.

The aliens involved in such cases were often claimed to be based on nearby planets. As technology advanced and allowed us to learn more about our solar system, it became obvious that the conditions on planets such as Mars and Venus were very different from those described by contactees. It has also been noted that as competition for media and popular attention between prominent contactees became more fierce, there was an evident tendency for each contactee's claims to become more elaborate. There was also a tendency for the date of the alleged first encounter to get pushed further back in time with each contactee wanting to claim that they were the first to be contacted. We can be certain that, whether the contactees were knowingly lying to their followers or genuinely deluded, their claims had no basis in fact whatsoever.

Close Encounters of the Fourth Kind

Although Hynek only proposed the three categories already described, more recent commentators felt the need to add a fourth category. Close encounters of the fourth kind are those alleged encounters that involve humans being abducted by aliens. One of the earliest of such cases was that of a Brazilian farmer, Antonio Villas Boas. He claimed that in 1957 he was abducted while working on his farm at night and forced to have sexual intercourse with an attractive female alien. The female alien allegedly made barking noises during the sexual act. This case was followed a few years later by probably the most famous alleged alien abduction ever, one which received worldwide media coverage. This was the case of Betty and Barney Hill, described earlier in Chapter 3.

The alleged abduction of the Hills in 1961 was taken much more seriously by the ufological community than any previous claims of alien contact had been. Many of the features of that case have cropped up regularly in subsequent ones, including an initial UFO sighting, so-called 'missing time' experiences, and the use of hypnotic regression to 'recover' full memories of the abduction. The case was the subject of a bestselling book by John Fuller, *The Interrupted Journey*. Public awareness of such claims was raised further by subsequent bestselling books, including *Communion* by Whitley Strieber, *Intruders* by Budd Hopkins, both published in 1987, and John Mack's *Abduction: Human Encounters with Aliens*, published in 1994. This last book was especially welcomed by the ufological community, because while Strieber was a writer of horror stories and Hopkins was an artist, John Mack was a Pulitzer Prize-winning professor of psychiatry at Harvard. So, to have a person of his stature proclaim that the abduction experiences reported in his book were 'not hallucinations, not dreams, but real experiences' lent much greater credibility to such claims.

It is unclear exactly how many people have claimed to have conscious memories of being abducted by aliens, but the figure is likely to run into many thousands. These accounts typically involve waking up in bed paralysed with a strong sense of presence, seeing aliens, being taken aboard a spaceship and subjected to various medical procedures, and being returned to bed. There are variations that involve being abducted from cars during long monotonous journeys, being given a tour of the alien spaceship or even taken for a ride, and being given messages to take back to humanity, typically warning of the dangers of pollution or nuclear war. We should be wary of 'one-size-fits-all' explanations of alien abduction claims given their richness and variety, but certain psychological factors do appear to be relevant to the vast majority of them, excluding the minority of cases that are simply deliberate hoaxes.

There is growing evidence to support the argument that most alien abduction claims are probably based upon false memories, that is, apparent memories for events that never actually happened at all. First, as a group, the personality profile of abductees is such that they appear to be particularly susceptible to false memories, compared with control groups. They score more highly on a number of personality measures that are known to correlate with susceptibility to false memories. These include fantasy proneness and hypnotic susceptibility, as well as dissociativity (the tendency to experience altered states of consciousness such as out-of-body experiences) and absorption (the tendency to become totally absorbed in one's own mental activity such as 'losing oneself' in works of fiction).

Second, in a study by Susan Clancy and colleagues at Harvard University, individuals reporting conscious memories of being abducted by aliens scored higher on a direct experimental measure of susceptibility to false memories compared with a control group. The measure used in this study was to present participants with a series of lists of words to remember. Within each list, each word was closely related in meaning to a specific word that was not itself presented. Thus, the words *snore, snooze, dream, blanket, bed, pillow* and *nightmare* might be presented, but not the word *sleep*. However, many people would then wrongly report that *sleep* was presented. The total number of such words incorrectly reported across all lists provides a measure of susceptibility to false memories.

Third, techniques used to allegedly 'recover' memories of alien abduction, such as hypnotic regression, are now widely recognised as being very likely to produce false memories based upon expectation, belief, fantasy, and fragments of real memories of films seen, books read, and so on. The apparent memories produced can feel very real, accompanied by vivid imagery and strong emotions. Such techniques are employed because it is widely believed within the ufological community that aliens are capable of wiping the memories of their victims for most of the abduction experience.

It is worth noting here that the use of dubious 'memory recovery' techniques such as hypnotic regression typically result in recovering exactly the type of memory that was expected. So, if the subject suspects that they were the victims of alien abduction, that is what their 'recovered' memories will confirm. If they believe that they were the victims of ritualised satanic abuse, that is what their 'recovered' memories will confirm. Likewise, they believe that they were Cleopatra or Napoleon in a past life. In all cases, the techniques used are identical and in all cases, in the absence of any independent evidence to support the claims, we should be extremely wary of accepting them as accurate accounts of events that really took place.

What kinds of experiences can lead people to suspect that they may have been the victims of alien abduction in the first place and so set out to recover full memories of the experience? These triggering experiences can be quite varied, including a possible sighting of a UFO, a 'missing time' experience, or the discovery of scars of unknown origin on the body, all of which are open to perfectly ordinary alternative explanations. For example, a 'missing time' experience may reflect nothing more than the common experience colloquially known as 'highway hypnosis', wherein, on a long and monotonous drive, one goes into a mildly altered state of consciousness and sense of time is altered. It is also a pretty safe bet that if you were to meticulously examine every square inch of your body, you would probably find marks and scars of unknown origin that you simply had never noticed before. The idea that they had been made by alien medical intervention is one of the least likely explanations.

The single most likely trigger of the belief that one may have been abducted by aliens, however, is one or more episodes of a phenomenon known as *sleep paralysis*. Sleep paralysis in its most basic form is very common. It is a period of temporary paralysis, typically lasting just a few seconds, that can occur when the sufferer is between sleep and wakefulness. It is a little bit disconcerting but nothing more. Between 10 and 30 per cent of the population report having had this experience at least once. A smaller proportion of the population, around 5 per cent, report additional symptoms that make the experience much more frightening and a smaller proportion again suffer from this more vivid form of sleep paralysis on a regular basis. These additional symptoms can include a very strong sense of an evil presence, visual hallucinations (such as lights moving around the room or monstrous figures), auditory hallucinations (e.g., voices, footsteps, mechanical sounds), tactile hallucinations (e.g., being held tightly or being dragged out of bed), pressure on the chest leading to difficulty breathing, and intense fear.

In broad terms, the causes of sleep paralysis are understood. The normal sleep cycle consists of both REM (rapid eye movement) and non-REM phases. REM phases are typically associated with vivid dreams. During REM sleep, the muscles of the body are actually paralysed, presumably to prevent the sleeper from acting out the actions of the dream. However, during an episode of sleep paralysis something goes awry and, to put it simply, it is as though the brain wakes up but the body does not. The result can be a terrifying episode during which the sufferer cannot move but is fully aware of his or her surroundings. In addition to this, dream imagery combines with waking consciousness. If the sufferer isn't aware of the phenomenon of sleep paralysis, they may worry that they are going crazy. If they then come across a book by a self-appointed UFO 'expert' stating that the very symptoms they have experienced are indicative of probable abduction by aliens, they are reassured that they are not losing their minds – and the next step is obvious. They must seek out the services of a hypnotherapist to help them to recover the full memory of the rest of their abduction experience. The end result is a detailed false memory of alien abduction.

Conclusion

Science cannot claim to have conclusively demonstrated that all of the many thousands of claims of close encounters with aliens are mistaken. But I hope that I have shown that plausible counter-explanations, based upon well-established psychological principles, exist for the various categories of 'close encounter' proposed by J. Allen Hynek. I would argue then that my fellow contributors to this book are correct when they state that we do not, as yet, know whether the Universe is teeming with life, or whether life evolved only upon our home planet. They will, I'm sure, be reassured by this conclusion and you, dear reader, can be reassured that you will not be wasting your time in reading and reflecting upon their chapters.

WHERE TO LOOK FOR LIFE ELSEWHERE

6

Home Sweet Home: What Makes a Planet Habitable?

Chris McKay

Planets galore. As a result of discoveries over the past twenty years, we now know that the Galaxy is teaming with planets, many of which, we assume, will be orbited by moons. Many of these are likely to be able, in theory, to support life. At the same time, our understanding of the worlds within our own solar system has deepened. We now anticipate the possibility of life on several worlds in the outer solar system – Earth and Mars are no longer the only places of interest to astrobiologists.

The discovery of life elsewhere in the Universe would have incalculably profound implications for us. If the life discovered is different from life on Earth and thus represents a 'second genesis', a second time that life has emerged spontaneously, then this will open up unique possibilities for comparative scientific study of another example of biochemistry. It will also be compelling evidence that life is common in the Universe. Knowing two separate examples of life immediately implies that there are virtually infinite examples in the Universe. This is the prediction of the 'zero–one–infinity' rule. This rule states that in many domains the only numbers that make sense are 0, 1 and infinity. The science fiction writer Isaac Asimov first applied this rule to the nature of the Universe in his classic novel *The Gods Themselves*.

The first step in searching for this second example of life is searching for habitable worlds. The criteria for a habitable world was initially based on Earth and centred on liquid water on the surface, warmed by a Sun-like star. But as we prepare to search for life elsewhere in the Universe, and considering the numerous extra-solar planets discovered, it is timely to consider the requirements for a habitable environment and enumerate the attributes and search methods for detecting habitable worlds and evidence of life.

Our understanding of life is, of necessity, based on studies of life on Earth. One might think that this would begin with a consideration of the question: What is life? However, a compact definition for life eludes us. Perhaps when we have many examples of life to compare we may craft such a definition but, more likely, it is in the nature of complex processes to be difficult to define. The second question that one might think of in the context of life elsewhere is: How did life begin? At present we can only say that life has been present on Earth for about 3.5 billion years, but where it started, how it started (Earth or elsewhere), and how long that process required are unknown.

Without a definition of life or a consensus theory for the origin of life, we can best proceed by considering the questions we *can* answer, such as: What does life need? What are the ecological limits of life? What is life made of? And, what does life do? The answers to these questions form the basis for understanding habitable worlds for life and how to search for evidence of life on them. Generally, the requirements for life on Earth can be compactly summarised as: energy, carbon, liquid water, and a few other elements.

Water

The fundamental ecological requirement for life on Earth is liquid water. Indeed, it is the availability of liquid water that appears to limit the occurrence of habitable environments on Earth and we assume this to be the case for life on other worlds. It is no surprise then that the search for life on other worlds is currently based on a strategy of 'follow the water'.

Europa, one of the large moons of Jupiter, is an ice-covered world without an atmosphere. However, there is convincing evidence that under the ice surface there is a global ocean warmed by tidal stresses as Europa orbits Jupiter. Pictures of Europa's surface returned by the *Galileo* spacecraft show features that appear to be icebergs and refrozen melt pods that indicate a subsurface ocean at some time. Furthermore, the magnetometer on *Galileo* indicated the presence of an ocean by detecting a global layer of slightly salty liquid water. The streaks on the surface of Europa may be cracks in the ice cover where perhaps ocean water has come to the surface.

Under the thick ice cover, the ocean water of Europa would be dark, isolated from external sources of organic matter, and presumably devoid of oxygen. Interestingly, there are a few microbial ecosystems on Earth that thrive in just these conditions.

Energy

Life requires energy to produce biomass and power reactions. On Earth, life obtains that energy from sunlight or chemical energy. Most ecosystems on Earth are powered either directly or indirectly by sunlight including subsurface ecosystems because most of them derive their energy from photosynthetically produced organic material that filters down from the surface. The microbial and animal communities found near deep-sea vents are sometimes cited as examples for how life could survive in Europa's subsurface ocean. However, these vent ecosystems derive their energy from the reaction of hydrogen sulphide emanating from the vent with oxygen dissolved in the ambient seawater, which originally came from surface photosynthesis, and thus sunlight.

But there are three microbial ecosystems known on Earth that do not need sunlight and are completely independent of oxygen or organic material produced by surface photosynthesis. Two of these anaerobic chemosynthetic ecosystems are based on methane-producing microorganisms that consume hydrogen derived from rock–water reactions in subsurface volcanic rocks, and the third is based on sulphur-reducing bacteria that use chemical energy, produced ultimately by radioactivity deep underground.

The main uncertainty with life on Europa is the question of its origin. Lacking a complete theory for the origin of life, and lacking any laboratory synthesis of life, we have to base our understanding of the origin of life on other planets by analogy with the Earth and assume that life on Earth originated on Earth. It has been suggested that hydrothermal vents may have been the site for the origin of life on Earth and in this case the prospects for life in a putative ocean on Europa look brighter.

Carbon and other elements

Carbon, and the associated set of molecules known as organic chemistry, is the stuff from which life is constructed. In addition to carbon, life on Earth uses a number of the elements, although this should not imply that all these elements are requirements for life elsewhere. Other than water and carbon, the elements nitrogen, sulphur and phosphorus are probably the leading candidates for the status of required elements. Life, in common with the Universe at large, has more hydrogen atoms than all other types of atoms put together. Taking *E. coli* bacteria as the model, 60 per cent of the atoms of life are hydrogen, 27 per cent are oxygen, 11 per cent are carbon, and 2 per cent are nitrogen. Other key elements, notably calcium, phosphorus, sulphur, sodium and chlorine, together come to less than 1 per cent of the atoms in *E. coli*. The dominance of hydrogen and oxygen, and their relative amounts, reflects the importance of H_2O in living systems. The four main elements of life – H, O, C and N – are among the most abundant elements in the solar system and in the Galaxy.

From these elements, life constructs a common core of biomolecules from which it can then build the great biochemical polymers it needs: the proteins, nucleic acids and polysaccharides. Proteins are composed of 20 amino acids, the nucleic acid DNA and RNA information-bearing molecules are made up from 5 nucleotide bases, and polysaccharides constructed from a few simple sugars. These polymers, together with a few distinct lipids (molecules such as fats, waxes and sterols), form the basic hardware of life. The software of life, the information stored in the genetic molecules, is also crucial to life and can be traced back to a single common ancestor.

It is worth restating this most distinctive property of biochemistry. Cosmically common elements have been formed by life into relatively simple molecules called monomers that can bind together to form more complex biomolecules. A good analogy might be common clay that is first formed into bricks, which are then used to build an elaborate mansion. Searching for these biochemical mansions, even if they are the remains of dead organisms, rather than searching for individuals that are actually alive is now the basis for the search for life in our solar system. It is thus instructive to look at the worlds of our solar system with habitability in mind.

What are the limits of life?

So, given liquid water, energy, and a few key elements outlined above, what else does life need to flourish on a planet? The answer is not much. Given suitable liquid water, life is robust and some life forms are able to tolerate high levels of ultraviolet and cosmic radiation. In fact, some photosynthetic organisms can use sunlight at levels thousands of times smaller than direct sunlight. Most of the limits of life are tied to the availability of water: at high temperatures water becomes less polar and the membranes of cells fall apart as a result. At low temperatures, of course, it becomes a solid. High salt and extremes of pH (acidity or alkalinity) can also be understood as alterations in the liquid water medium in which life operations take place. When these alterations become too extreme, life does not thrive. The species of cyanobacteria found in the salt domes of the hyper-arid region of the Atacama Desert grow in the presence of water stress, oxidants and high salinity. They represent Earth's entry into the record book of life forms capable of growing in extremes.

What does life do?

All the many activities of living systems can be usefully summarised as Darwinian evolution: the continuing cycle of reproduction, mutation and selection. This is what life does and is what separates life from complex, open, but non-living systems, such as hurricanes, which are similarly born, go through a life cycle (during which they self-organise and consume energy), and die. Such non-living systems can even reproduce. Why are they then not considered life? The answer is that they do not have information stored in genetic material. That information is what allows Darwinian evolution and hence is effectively the difference between life and self-organising open systems. We assume that life elsewhere will also be characterised by Darwinian evolution, even if it is instantiated in radically different molecules, and even different elements, than the life we know.

Could life exist without water?

The search for life typically assumes that it requires a liquid that must be water, or at least water-based solutions. However, Titan, the largest moon of the planet Saturn, has a thick atmosphere composed primarily of nitrogen and methane with many other organic molecules present. The surface pressure is 1.5 times the sea level pressure on Earth. The temperature at the surface is close to −180°C, which is much too cold for water to be a liquid, but cold enough to liquefy the methane in the atmosphere. Titan presents us with a challenge to our assumption that life requires liquid water. If we consider the possibility of life based on liquid methane and ethane then all of our background assumptions about biochemistry must be re-examined, because the biochemistry of life on Earth is finely tuned to the properties of water.

The organic material in Titan's atmosphere provides a potential source of chemical energy for life and the liquid methane on the surface could provide a possible liquid medium for an alternative type of life. Liquid methane is cold and not a good solvent compared with water. To live in such a dilute solution would require some method to actively seek out nutrients and bring them into the cell. Such cells might be shaped like large sheets of paper to maximise the surface area through which nutrient can be collected. Enzymes could catalyse the necessary reactions even at low temperatures. If carbon-based life in liquid methane exists on Titan, it could be widespread and have global effects on the atmosphere. The most probable chemical energy source available for life on Titan involves the consumption of hydrogen, probably together with acetylene. Hence a depletion of hydrogen in the atmosphere near the surface may be the most easily observable effect of life on Titan.

Based on what we know about the physics and chemistry of the environment on Titan we might expect that if there is life, then the small diversity of elements available in the environment would limit its complexity and its ecosystems. This would be exacerbated by the low temperature of the liquid solutions, resulting in negligible solubility.

Given these limitations it may be that if there is life in the liquids on Titan's surface it may be simple, heterotrophic (unable to generate its own food as an Earth-based plant could), slow to metabolise and slow to adapt with limited genetic and metabolic complexity. Because liquid methane and ethane are widespread on the surface of Titan, the simple molecules needed for metabolism may be widespread in its environment, but the complex organics needed for structural or genetic systems may be hard to obtain or synthesise. The communities formed may be ecologically simple – perhaps analogous to the microbial ecosystems found in extreme cold and dry environments on Earth.

The advantages of the Titan environment for life include free food from the sky, in the form of organics (principally acetylene and hydrogen), the chemically benign nature of the non-polar solvent in terms of degrading biomolecules (in contrast to water), lack of ultraviolet or ionising radiation at the surface, and low rates of thermal decomposition due to the low temperature. Titan may have only a simple trophic system, probably without primary producers and without predators. Photosynthesis may be beyond the complexity that can be achieved with the limited elemental, and hence genetic, diversity – but food is free. The simple low-temperature life forms and communities envisaged would have very low energy demands and would grow slowly. Life on Titan may be simple, but if it has a genetic system and is thus truly Darwinian then it would be a clear and fascinating example of a second genesis of life.

Exoplanets

The discovery of exoplanets and exomoons is expanding rapidly and it is clear that we will find many worlds that resemble the Earth – worlds more Earth-like than any of the other worlds of our solar system. Considerations of the habitability of exoplanets would follow the logic applied in studying our own solar system. Thus, the requirements for life on Earth, the elemental composition of life, and the environmental limits of life, provide a way to assess the habitability of exoplanets and exomoons. Temperature is key both because of its influence on liquid water and because it can be directly estimated from astronomical observations and climate models of exoplanetary systems. Life can grow and reproduce at temperatures as low as −15°C, and as high as 122°C. Studies of life in extreme deserts on Earth show that, on a dry world, even a small amount of rain, fog, snow, and even atmospheric humidity can be adequate for photosynthetic production, generating a small but detectable microbial community. Life is able to use light at levels less than one part in a hundred thousand of the solar flux at Earth. Ultraviolet and cosmic radiation can be tolerated by many microorganisms at very high levels and is unlikely to be life-limiting on an exoplanet. Levels of oxygen over a few per cent on an exoplanet would be consistent with the presence of multicellular organisms, and high levels of oxygen on Earth-like worlds indicate photosynthesis similar to that in green plants, providing the conditions necessary for large animal life. Other factors, such as pH and salinity, are likely to vary and therefore not likely to limit life over an entire planet or moon.

We may soon discover numerous Earth-like exoplanets and find conclusive evidence of biologically produced gases (oxygen, methane, etc.). However, there is no known way to follow up with an astrobiological investigation into the nature of the life present, and thus to determine its biochemistry and therefore whether it is truly a second genesis or related in some way to Earth life. Unlike the worlds of the solar system, the biochemical investigation of life on a distant extrasolar Earth may be many human lifetimes in the future.

What if we succeed?

In closing, it is interesting to ponder the implications of finding life nearby, say on Mars. If we do discover a second genesis of life there, does it have any ethical implications or is it purely of scientific interest? Current international rules for planetary protection focus on protecting future scientific investigations rather than protecting extraterrestrial organisms or ecosystems. If we discovered a second genesis of life on Mars – even if the representatives of that life were only microscopic – this discovery would raise new and profound issues in environmental ethics and would, or at least should, cause us to think about how we act with regard to that life. We will need to consider what ethical considerations are necessary for an alien life form when that life is distinctly different from Earth life and the members of that life are no more advanced than microorganisms. I hope we have the wisdom to shift the focus of our space exploration from merely harvesting scientific data to protecting and enhancing the richness and diversity of life in the Universe.

7

The Next-Door Neighbours: The Search for Life on Mars

Monica Grady

In any book about aliens, there has to be a chapter about Mars. This is because – as our literature and our movies show – we are fascinated by Martians, especially if they are malevolent and hostile to terrestrial beings. But what is the true likelihood of life on Mars? Speculation has been rife for hundreds of years: the astronomer Giovanni Schiaparelli's map of 1877, drawn from his telescope observations and showing details of *canelli* ('channels'), was thought by many to provide evidence of a Martian civilisation. But we know now that the features Schiaparelli detailed were artefacts, and not channels carrying water as he had thought. And it is widely supposed that mistranslation of the italian *canelli* to canals, rather than channels, was responsible for propagating the belief that the features were constructed by Martians. Images of Mars's surface from orbiting spacecraft and rovers on the ground have provided us with detailed pictures of a dry and dusty landscape, with no sign of anything living. No lichens cover the rocks, no stains from algae decorate gully walls, and no twisted remnants of scrub cling to desert outcrops. Mars appears to be barren – but still we send spacecraft there to search for life. Why? What makes us think that there might be something living there? Let's review what we know about the planet, and consider its biological potential.

Introduction to Mars: How does it differ from Earth?

Mars is a rocky planet with a diameter around half that of the Earth. It is half as far away again from the Sun as the Earth, and takes almost two Earth years to orbit it. What is slightly strange about Mars is that it rotates on its axis at a very similar rate to the Earth, and so a Martian day is only very slightly longer than a day on Earth. It has a very thin atmosphere, mostly of carbon dioxide, with a pressure of only around 6 mbar, compared with average terrestrial air pressure of approximately 1000 mbar. One of the many benefits of our atmosphere is that it insulates the planet – so we have a global mean surface temperature (GMST) of around +15°C. Mars has no such insulating blanket, and so the warmth of the Sun (which is less than half that on the Earth because Mars is further away) is reduced greatly: the GMST is −55°C. For comparison, if the Earth lost its atmosphere, its GMST would be around −18°C.

As well as insulating against the cold, our atmosphere also protects us against cosmic and solar radiation. Galactic cosmic rays, highly energetic particles travelling at around the speed of light, spread through the solar system. Interaction between the particles and Earth's atmosphere prevents most of the cosmic rays reaching the surface. Similarly, about three quarters of the harmful ultraviolet radiation from the Sun is absorbed by the atmosphere. In contrast, the surface of Mars receives a much higher radiation dose than is healthy – over seventy times that of Earth each year.

Like the Earth, Mars started out as a planet with a surface kept molten by asteroid and comet bombardment and a core heated and melted by the decay of radioactive isotopes. At one stage, Earth and Mars shared a similar internal structure of a metal core overlain by a rocky mantle and crust. But because Mars is only about half the size of the Earth, its heat dissipated more rapidly and the planet is now completely solid, while Earth is still molten around the core. The rotation of the molten core leads to generation of a magnetic field around Earth – we can see this every time we look at a compass needle. The magnetic field is another defence mechanism against galactic cosmic rays as well as energetic particles from the Sun, deflecting them away from the Earth. Without the magnetic field, we would, as on Mars, be subject to an increased radiation hazard.

Just as the presence of a magnetic field on Earth follows from having an active core, the occurrence of water on Earth's surface follows from our planet's thick atmosphere – and Mars's lack of atmosphere helps to explain its desert landscape. At sea level on Earth, water boils at 100°C. As you go up a mountain, the boiling point of water drops by around 1°C for every 300 metres you scale. This is because, as the air pressure decreases, the boiling point of water decreases – so on Mars, with a surface pressure of only 6 mbar, any water evaporates instantly – H_2O is not stable in liquid form. It is present as ice, both buried under the surface and as ice caps at the poles, but rivers, streams, lakes and oceans are absent on Mars.

The final major feature that sets Mars apart from Earth is plate tectonics; or, in Mars's case, the lack thereof. We have a very dynamic Earth, with its activity powered by the underlying rotation of the molten core. Plate tectonics is a vital part of both the carbon and water cycles that move volatiles between different reservoirs. For example, the leaves on a tree fix water and carbon from the atmosphere. The leaves die, then rot to become soil. Over millions of years the soil is transformed to rock forming part of a tectonic plate, which may be subducted (pulled down into the deeper crust) and melted. Carbon and water return to the atmosphere as volatiles during eruptions of molten rock from a volcano. Without such cycles the Earth would stagnate – just as Mars, without its underlying dynamo, seems to be stagnating.

So Mars differs from Earth in its internal structure, its insignificant magnetosphere, its lack of atmosphere and hydrosphere, and the absence of plate tectonics. All these features follow from the difference in size of Earth and Mars. So why might we still nurture a hope that Mars harbours life?

Why could there be life on Mars?

The previous section painted a very dismal picture of Mars as a cold, dry and inert planet with a surface bathed in radiation – not the sort of environment conducive to a flourishing ecosystem. But Mars has not always been like this – when the planet first formed, it could have been a haven for life, with flowing rivers and inland seas, and a rich diversity of potential habitats warmed by heat flowing from its molten interior. To understand the dramatic change in its planetary circumstances, we have to look at how Mars has evolved since its initial formation.

Some 4.6 billion years ago, the solar system formed from a turbulent disc of gas and dust that, over the course of about 3 million years, formed into the planets and satellites that we have today. Earth and Mars were produced from the same ingredients, by the same mechanism, in neighbouring locations in the disc, differing only by radius. As outlined above, it is the disparity in size between Earth and Mars that has led to differences in cooling rate, in turn influencing every aspect of the development of both planets' solid, liquid and gaseous reservoirs. Despite this, there have been periods in Mars's early history when life might have got going.

The first essential requirement for life is to have the correct ingredients. The building blocks of life are the molecules of hydrogen, carbon monoxide and ammonia (H_2, CO and NH_3). These ingredients were plentiful during the formation of the Universe, and so must have been present in the planet as it formed. The second essential requirement for life is water – or some kind of fluid – which provides a medium for concentration and transport of molecules, allowing them to react together. There is a huge volume of evidence showing that water was abundant on the surface of Mars in its earliest days. Orbiting satellites have taken pictures of the landscape, and images show the patterns left on the surface by rivers, streams, lakes, deltas and inland seas. Spacecraft have mapped the distribution across Mars of different generations of rocks containing clay minerals – sediments that must have been deposited from bodies of standing water. On the ground, we have seen cross-bedded and stratified rocks, water-worn pebbles and gravel in pictures taken in close-up from cameras on landers and rovers. It is hard to dispute that Mars has had a significant fluvial history – which implies that, for many millions of years, water was stable on its surface.

The third essential for life to arise is an environment in which complex molecules can survive without being broken down – so one with a reasonable temperature and low levels of radiation. The evidence for abundant water flowing over the surface of Mars indicates that sometime in its past, the atmosphere was sufficiently thick to allow water to exist as a liquid for extended periods of time. With a thicker atmosphere, the surface would be warmer and protected from radiation.

When Mars first formed, the planet had all the ingredients for life to get going, water to facilitate chemical reactions, and an environment suitable for living organisms to replicate and grow. Although tectonic activity in the form of plate movement was absent, there was still sufficient heat in the cooling interior of the planet to fuel active volcanism for many millions of years, continuing the build up of the atmosphere. On Earth, the environment, although changing (from an oxygen-poor atmosphere to one that was oxygen-rich), was still conducive for life, and trace fossils have been discovered with ages as old as 3.8 billion years. So it is entirely feasible that, say, 4 billion years ago, the same type of microbial life could have developed on Mars as was starting to develop on Earth. Gradually, though, things changed on Mars: there was a gradual loss of chemicals with low boiling points, such as carbon dioxide and water, as the planet cooled and the volcanoes that would have returned them to the atmosphere went extinct, as well as, most likely, a more catastrophic loss of a large percentage of the atmosphere through solar wind sputtering. This is a process whereby particles in the solar wind 'tear' the atmosphere away from the planet – Mars's smaller size means that it has a smaller gravitational attraction than Earth, and it is much easier for the atmosphere to be removed in this way. By about 3.5 billion years ago, almost all of Mars's atmosphere had gone and, with it, the chance of the evolution from microbial organisms to major surface-dwelling species.

If there was life on Mars, where might it have arisen? And where might it be now?

Given then that during Mars's earliest history the same conditions existed there as on Earth, logically there should be no reason why life could not have developed. In which case, since Mars's climate was much more clement up until about 3.5 billion years ago, where might microbial life have got going? One flippant answer is 'almost anywhere', but we can be confident that any surface-based life form would have become extinct once the atmosphere (and water) disappeared, and the surface was subject to elevated levels of radiation.

Looking to where life might have survived leads us to search both inwards – to the interior of rocks – and downwards to what is available below the surface. On Earth, there is a large group of organisms called endoliths (literally: 'inside rock'). This group is made up of chasmoendoliths, which live in cracks or holes in a rock, remaining in direct contact with the external environment; and cryptoendoliths, which infiltrate rocks, colonising between mineral grains and within pores, where they are protected to some extent from direct external influence. A third type of endolith (euendolith) can bore or tunnel quite deeply into rocks. Endoliths are not a single type of organism – the term simply describes the habitat in which they survive. Endoliths can be organisms from all the domains of life: the unicellular archaea or bacteria, as well as the multicellular eukarya. They may be a single species, or a symbiotic relationship between species. For example, in Antarctica, there is an enormous, almost invisible, biomass of cryptoendoliths that inhabit the sandstones that comprise much of the exposed rock layers of the continent. The microorganisms can be seen as narrow (submillimetre) layers a few millimetres below, and parallel with, the outer surface of the rocks, protected from the cold and the wind, deriving nutrients from the rock as well as by photosynthesis. One of the layers comprises of cyanobacteria, a second is a fungus. It is not difficult to imagine the survival of such a colony on Mars – but, so far, examination of the subsurface layers of rocks by the Curiosity rover has not revealed any indication of such specimens.

Penetrating more deeply below the Earth's surface in recent years has revealed a flourishing ecosystem within caves. Shallow caves are inhabited by higher species, such as bats, that obtain their food from the surface, and by other animals that exist on material that is carried into caves by streams or during flooding. Clearly, these are not the troglodytes that might exist in caves on Mars. Deeper below the terrestrial surface, cave dwellers are limited to microbes, able to survive because most cave systems are wet, with streams and pools of water available for colonisation. Even 'dry' caves contain some water, even if only held within the pore spaces of the surrounding rock. The microorganisms that inhabit cave systems employ chemosynthesis – whereby energy is derived from oxidation–reduction reactions between inorganic molecules – for survival, rather than direct photosynthesis, which requires sunlight. Such microorganisms still require oxygen, which on Earth is produced by photosynthesis. They derive other nutrients almost entirely from the rocks, and often produce methane gas as a waste product. On Mars, any cave dwellers may also survive through chemosynthesis, possibly using carbon monoxide instead of oxygen as the oxidiser in a metabolic pathway. Although caves have been identified in images of Mars's surface, no cave system has yet been explored to determine whether there is a subterranean ecosystem.

Is there any evidence for life on Mars? What evidence should we look for?

To date, none of the spacecraft orbiting or present on Mars has observed anything that looks like life. The two Viking landers of 1976–77 had on-board experiments to test for biological activity, but results from the instruments were ambiguous. This is partly because the samples they took were from the surface of the planet, and solar ultraviolet radiation is likely to have removed any organic compounds present in the top few millimetres of the soil.

If there are no organic compounds at the surface, are there any deeper down? The Curiosity rover has drilled to a depth of 7 centimetres, which is probably still insufficiently deep to be below the zone of radiation interaction. Even so, organic molecules have been detected – although there are still some concerns about whether or not they are a result of contamination. An alternative source of information about Martian organics comes from analysis of Martian meteorites. But here, too, although organic species have been detected, the possibility of terrestrial contamination has not been completely ruled out. So the existence of indigenous organic materials on Mars is still unconfirmed.

What other signs of life should we be looking for? On Earth, fossils of increasing complexity have allowed us to trace the evolution of plant and animal life from simple microorganisms up to multicellular, highly specialised varieties. On Mars, searching for fossil evidence of life is well-nigh impossible: although the currently active rovers, Curiosity and Opportunity, are examining the Martian surface and near subsurface with a diverse array of instruments, including microscopes, neither rover is equipped to split rocks in a way that would reveal traces of fossilised microbial life. There was one report of a fossil found in a Martian rock – a possible petrified bacterium was identified in a meteorite in 1996. However, the claim was controversial, and twenty years later, there is still debate about the discovery. Part of the problem springs from the fact that the meteorite was in Antarctica for about 13,000 years prior to its recovery, so there was plenty of time for terrestrial microbes to have colonised the specimen.

There is one signal that is regarded as being diagnostic of the presence of life – and that is the occurrence of methane in a planet's atmosphere. Methane is destroyed by ultraviolet radiation, so its presence implies an active source that continuously replenishes the atmosphere. The source may be abiotic – for example, the weathering of silicate rocks, such as basalts, generates methane, as does melting of permafrost on Earth. Methane may also have a biological source: on Earth, termites and ruminants are major methane producers – the gas is generated internally by bacteria living in their digestive systems. Indeed, microorganisms in a variety of habitats are the main source of methane on Earth.

Methane has been found in Mars's atmosphere by three separate techniques. It was first detected by Earth-based telescopes, where it seemed to be concentrated into clouds or plumes over specific regions of the Martian surface. When the observations were repeated three years later, the methane had disappeared, leading the observers to propose that the clouds were seasonal, and only occurred in the Martian summer. Methane was also detected by an instrument on board the *Mars Express* orbiting spacecraft, again seen to be concentrated in clouds but over different regions from where the telescope observations found methane. Unfortunately, the interpretation of both sets of results (telescope and spacecraft) is not straightforward, and so it's hard to draw any specific conclusions about what the level of methane means. As yet, there is no definitive evidence for a biological origin of the methane. So if methane-producing bacteria are living in subsurface ecosystems on Mars, they are not generating sufficient methane to be detectable.

Are there any aliens on Mars?

The short answer to this question is yes, especially if we use the dictionary definition of *alien* as 'belonging to something else'. There are at least eight aliens on Mars, six of which are silent and motionless, and two, named Curiosity and Opportunity, which are still exploring their surroundings, stopping every now and again to prod a rock or photograph the scenery. They are the true aliens – artificial intelligences following the commands of a remote controller on a distant planet. This chapter, though, is not about 'aliens' on Mars – it's about Martians on Mars. Are there any? So far, we have not found any evidence of life – but we haven't yet explored underground or dug or drilled sufficiently deep below the surface to know for sure. The jury, as they say, is still out. Maybe there are Martians on Mars – assuming, that is, our own 'aliens' haven't killed them all off in the search for evidence of their existence.

8

Further Out: Could the Moons of the Gas Giants Harbour Life?

Louisa Preston

A source of liquid water, energy and nutrients are the classic conditions we currently understand to be needed for life. From our experience observing Earth's diverse biota, wherever there is water, there is a good chance of finding living organisms. This is one reason why the growing number of inferences, and indeed sightings, of liquid water pockets across the solar system is driving the search for extraterrestrial life. However, while Earth is (mostly) an extremely hospitable environment for life, conditions on other planets can be far less favourable. But what gives us further hope of success is the discovery that severe environmental conditions on Earth, within which certain life forms can thrive, are similar to physical and chemical conditions found to exist on both planets and moons in our solar system and beyond.

This 'extreme-loving' life sits under the umbrella term of the *extremophiles* and can be in the form of unicellular organisms, such as bacteria and archaea, to multicellular organisms, including penguins and tardigrades (the water bears). These last are tiny translucent animals less than half a millimetre in size that resemble armoured eight-legged pandas. They are found just about everywhere on Earth, having survived all five mass extinctions and even endured the radiation-rich vacuum of space. Specific types of extremophiles that are useful in the search for extraterrestrial life are the thermophiles (heat lovers), psychrophiles (cold lovers), halophiles (salt lovers), barophiles (living under high pressures), acidophiles (living in low pH, or acidic, conditions), alkaliphiles (living at the higher range of the pH scale), anaerobes (living without oxygen), and those that can withstand extreme radiation. Each and every one of these organisms has the potential to be able to withstand the harsh environments existing on other worlds in the solar system – though whether they actually do or not is still a question to be answered.

Although incredibly diverse, today we have but one example of life, and a single template of a world capable of supporting it. The planet Earth is perfect for life – a rocky world with a substantial protective atmosphere that sits within the habitable, or 'Goldilocks', zone of the solar system, allowing it (for the most part) to remain at a temperature suitable for liquid water, and hence life. Despite other planets being the obvious place to look for alien life, it could just as easily be sustained on a moon and there are some key environments which make particular moons of the solar system of great interest in the search for extraterrestrial life. Within the Goldilocks zone the only candidate satellites are the Moon itself and the Martian moons of Phobos and Deimos, but none of these has an atmosphere or water in liquid form and as such are not serious candidates for evidence of life. However, the Goldilocks zone is not only larger than previously thought, it can also be found in more than one place in the solar system. It is now believed there are multiple Goldilocks zones which, instead of forming habitable bands around the Sun, circle the outer planets. This points the search for extraterrestrial life in the direction of these moons and there are more than 210 frozen natural satellites orbiting the gas giant worlds of Jupiter and Saturn to choose from. Indeed, a few of these moons are proving to be astrobiologically very interesting – much more so than the planets they orbit!

Astronomical observations and recent space missions have revealed many of these outer solar system moons to be geologically active bodies, with volcanoes spewing ice as well as molten lava, geysers the size of whole countries on Earth, impact craters in their thousands, and vast channels and valley networks. Most excitingly, these moons are also displaying a wealth of potentially habitable environments. Their frozen surfaces may not be the most suitable for life, but the question of whether liquid water oceans are hidden beneath their icy shells, which may have existed long enough to have been biologically useful, has motivated the search. If liquid water exists below ice layers and is in contact with heat sources radiating from the interior of the moons (via radioactive decay, volcanic interactions or hydrothermal activity) then these moons may be considered as potential habitats for life. The problem for astrobiologists, however, is that although a number of geological features on these icy moons could be housing alien beings, they are so incredibly extreme compared to some of the harshest environments on Earth that suitable terrestrial comparisons or *analogues* of the environments – and the potential life forms – are fewer in number and harder to find. For now, much of our knowledge of the conditions actually present on these icy moons is based on inferences rather than on definitive data. There is a lot more educated guesswork and imagination needed for finding life on these worlds than those nearer to home – but then that is half the fun.

The gas giant realms

While there have been no samples taken that could actually test for microscopic life on Jupiter or Saturn, there is quite a bit of compelling evidence that shows there is no possible way for life as we know it to exist on them. Composed mainly of hydrogen and helium, there is virtually no water to support life on Jupiter. The planet does not have a solid surface for life to develop on, so the only tiny possibility for life would be in a floating microscopic realm high up in the atmosphere. However, even this seems unlikely as the atmosphere of Jupiter is in constant chaos. So even if life somehow could hold on in the lower-pressure upper reaches, and could resist the harsh solar radiation there, it would eventually be sucked down into realms where the pressure is thousands of times that of Earth's atmosphere and where temperatures rise to over 10,000°C. It would be almost instantaneously destroyed. No life form on Earth we currently know of could survive in anything close to such environments.

If life is near impossible on Jupiter, you can guarantee the same can be said for Saturn. Like its larger neighbour, it too is composed almost entirely of hydrogen and helium, with only trace amount of water ice in its lower cloud deck, and it too has no surface upon which life could live. At the top of the clouds the temperatures are around −150°C, and although it gets warmer as you descend through the atmosphere, the pressures increase too. Sadly, once temperatures are warm enough for liquid water, the pressures are simply too high for life. It is also pretty windy up there, with speeds up to 1000 mph.

Europa

One of the most important and interesting moons in the search for habitable environments in the solar system is Europa. At first glance Europa doesn't look particularly appealing for life: although formed from silicate rock like the Earth and the other terrestrial planets, instead of a watery liquid it is covered in a smooth 62-mile-thick sheet of water ice. It is constantly bombarded with ionising radiation as it lies within Jupiter's magnetosphere and temperatures at its surface range from −187 to −141°C, far below the lowest limits for microbial growth, which is not surprising since it's about half a billion miles from the Sun (more than five times further out than Earth).

The surface ice itself is not an environment that any currently known terrestrial life form could withstand. However, the ice could provide just enough protection from the intense bombardment of radiation to allow for the preservation of organics and even organisms beneath it, as well as encourage more favourable temperatures. Just as a layer of ice over a pond insulates the water beneath it, allowing it to stay liquid and aquatic life to go on living, Europa's rind of ice shields its vast ocean, helping it to keep warm enough to remain fluid despite the moon's great distance from the Sun. As Europa orbits around Jupiter it is bowed and stretched by its massive parent planet, generating a heat deep inside that also helps to keep its water from freezing. You might think that this sheltered salty ocean is pretty small, but in fact Europa is only slightly smaller than the Moon, and the volume of its ocean is estimated to be 3×10^{18} cubic metres – twice the volume of all Earth's oceans put together. Potentially active volcanoes and vents may exist at the base of the ocean, which act to heat the water and provide sites where bacterial life could thrive, as it does on Earth. Thus we see that Europa has two key elements thought needed for the development and even persistence of life: water and heat energy. We just need to find some organic chemicals.

Europa provides an extreme environmental challenge for life, but a number of extremophilic organisms and icy cold habitats on Earth could offer analogues for liveable environments there. Firstly, there is an important similarity between salty lakes on Earth and the Europan ocean itself. Lake Tirez in Spain contains very salty sulphate-rich waters that may be similar to Europa's briny interior. In addition, salt-loving halophiles are found to grow and thrive in these waters. Secondly, Earth has a number of liquid water lakes hidden under ice, some found up to two miles beneath ice sheets in Antarctica, which are kept liquid by a combination of geothermal heat and the pressure of the overlying ice. Lakes Vostok, Ellsworth, Bonney and Vida are thought to be similar to Europa's salty subsurface ocean and in addition display the capabilities of life to survive for millions of years buried under a shield of ice. Cores taken from the ice above the largest known *subglacial* Lake, Vostok, in 2012 revealed DNA from an estimated 3507 organisms. A habitable sea-floor environment may also exist on Europa. There is a range of extremophile communities in the dark, cold, high-pressure environment of the Earth's ocean floor – particularly around deep-sea hydrothermal vent fields, such as Lost City on the Mid-Atlantic Ridge and the Mariana Trench in the Pacific Ocean. These analogues are important to investigate, even if at present any physical search for a deep-ocean biosphere on Europa is impossible.

Enceladus

Saturn's sixth-largest moon, Enceladus, is 314 miles wide and covered by an icy shell with a surface temperature at noon of a chilly −198°C. When NASA's *Cassini* probe flew by in 2005 it piqued the interest of astrobiologists with the sighting of present-day geological activity occurring at its surface. Gargantuan jets of fine icy particles and water vapour were observed erupting from cryovolcanoes at its south pole, which is surprisingly warm considering it is made of ice. Over a hundred of these jets have been seen so far, culminating in a large plume that soars several thousand kilometres into space and contains not only water, which scientists think originates from a subsurface ocean, but also simple organic carbon-based molecules and volatiles such as nitrogen, carbon dioxide and methane (N_2, CO_2, CH_4) – similar to the chemical make-up of comets. Each individual exploding geyser rises over 400 kilometres into space (about the distance from London to Paris) and while some of the water vapour spewed out falls back onto Enceladus as snow, the rest escapes and supplies most of the material making up one of Saturn's rings (called the E-ring).

Data gathered during Cassini's fly-bys suggests that underneath this frozen outer shell lies a rocky core with a global water ocean sandwiched in between. It is thought that this ocean is in physical contact with Enceladus's rocky mantle, which means that a whole host of interesting chemical reactions useful to life are possible. The ocean is likely to be full of sodium chloride (common table salt), as is found within Earth's oceans, and is alkaline with a pH of 11 or 12. This alkalinity may be caused by interactions between metallic rocks and the water, resulting in the production of molecular hydrogen (which is also a source of chemical power). These chemical reactions deep within Enceladus have the potential to free up energy that could be used to support a subsurface biosphere. This alien water body is also full of carbonate-rich salts, so is probably more akin to soda lakes such as Mono Lake in California than it is to the Atlantic or Pacific oceans. A variety of extremophilic organisms survive in this type of salty terrestrial realm. Active hydrothermal vents are thought to exist on its sea floor, suggesting that conditions there could be similar to those that gave rise to some of the first life forms on Earth. Compounds essential for life must have come from sources deep inside the moon, from the place that feeds the jets. The assumption we make, therefore, is that organic molecules used by, and needed for, life are present deep within Enceladus.

A plausible subsurface ecosystem on Enceladus would be unlike many terrestrial ones as life forms would have to be independent of oxygen and not rely on organic materials produced by the reactions of photosynthesis (the process of using energy from sunlight to make biomolecules from carbon dioxide and water). The young volcanic landscapes of Iceland provide a good analogue for these plumes, and indeed for life forms. Iceland is covered in geysers and hot springs, cracks in the Earth's surface where near-boiling water erupts in spectacular fountains, blanketing the ground with mineral and nutrient-rich waters. Surrounding the hot springs of Iceland are heat- and acid-loving bacteria forming mats of microorganisms, creeping across the surface, thriving in the hot acidic waters. Such features are minuscule versions of the colossal jets seen erupting from Enceladus; however, they can inform us about the processes involved in their formation and their ability to create and support a habitable environment.

Titan

One place that, in appearance at least, greatly resembles the Earth is one of Saturn's haziest moons, Titan. Deceptively Earth-like, Titan has a dense nitrogen-rich atmosphere complete with clouds and seasonal storms which leave wet patches on the surface that are large enough to be visible from orbit. Sunlight and electrons stream across Titan from Saturn's magnetosphere and break apart the nitrogen and methane in its atmosphere, starting a cascade of reactions that end with the production of carbon-rich compounds – this creates a solid organic haze that fills the atmosphere and shrouds the surface from view. It has a very familiar landscape beneath this seemingly impenetrable veil with mountains, dunes, riverbeds, lakes, shorelines and seas. One lake, Kraken Mare, is three times larger than the Great Lake Michigan–Huron.

That is where the similarity ends. The Cassini spacecraft revealed a world on Titan that may look very much like ours but has a completely different chemistry. Surface temperatures of around −179°C have meant that the liquid bodies on the surface could not possibly be composed of water, but are more likely a mixture of methane and ethane. These hydrocarbons (that on Earth are found as gasses) are able to flow as liquids across the surface due to Titan's chilling environment. In fact, the volume of liquid hydrocarbons resting in Titan's second largest sea, Ligeia Mare, is a hundred times more than all the oil and gas reserves on Earth combined, and are most likely formed by some very different processes. That is not all that is eerily similar yet drastically different. Instead of rocks, Titan is littered in blocks of water ice. Instead of molten lava, it has a slush of water ice mixed with ammonia. Instead of surface dirt, it has smog particles that have rained out of the sky, and it even has wind-blown dunes, which may be made of organic hydrocarbons instead of sand.

Titan would probably be the most promising place, rather than Enceladus or Europa, to look for extraterrestrial life in the solar system if not for its frigidity. At Titanian surface temperatures, phospholipids – the chemical compounds that make up cell membranes – and the water found within all cells would be frozen solid. Any life that evolved on Titan's surface would therefore have to be made of a very different set of chemicals from those that constitute Earth-based life, and not be reliant on water, which is locked in a state inaccessible for it. Even so, life might be present within a range of unusual habitats, from the liquid hydrocarbon lakes on the surface to great depths within the subsurface, and potentially even in deep subsurface oceans of water and ammonia like that found within Europa and Enceladus – creating a potential biosphere volume double that of the Earth.

Owing to the very different chemistries of the liquids on Titan, however, we can only speculate as to what life there might be like, and as such there are very few analogues on the Earth for this world. The hydrocarbon lakes are crucial targets for the search for habitable environments and life on Titan. The best-known analogue for these is Pitch Lake, on the island of Trinidad. It is a natural liquid hydro-carbon lake just like those found on Titan, albeit a much smaller version. A unique extremophilic microbial community is found here, one that includes archaea and bacteria that live without any oxygen. The natural asphalt-soil seeps of the Rancho La Brea Tar Pits in California and the Alaskan Oil Field petroleum reservoirs are also potential habitable analogue sites for Titan. Despite its many differences, the prebiotic conditions of the environment on Titan and the associated organic chemistry are, in theory, enough to start a chemical evolution analogous to what is thought to have kick-started life on Earth. As such, scientists are very excited about Titan.

The future

Many scientists looking for life away from the Earth have a very different idea of what these life forms might look like compared with the aliens of popular culture. They are not looking for, nor realistically expecting to find, little grey or green men, but are more likely to detect simple extreme-loving microorganisms, organic compounds such as amino acids, and biosignatures indicative of past life. The more we learn about the extremophiles and the extreme environmental conditions they can withstand on Earth, the more plausible life on other worlds, in particular the moons of the solar system's outer planets, becomes.

9

Monsters, Victims, Friends: Aliens in Science Fiction Writing

Ian Stewart

If it is just us, seems like an awful waste of space.

— Carl Sagan, *Contact*

'On and on Coeurl prowled ... Jagged black rock and a black, lifeless plain took form around him. A pale red sun peered above the grotesque horizon ... His great forelegs twitched with a shuddering movement that arched every razor-sharp claw. The thick tentacles that grew from his shoulders undulated tautly. He twisted his great cat head from side to side, while the hair-like tendrils that formed each ear vibrated frantically, testing every vagrant breeze, every throb in the ether.

'There was no response. He felt no swift tingling along his intricate nervous system. There was no suggestion anywhere of the presence of the id creatures, his only source of food on this desolate planet. Hopelessly, Coeurl crouched, an enormous catlike figure silhouetted against the dim, reddish sky line, like a distorted etching of a black tiger in a shadow world.'

In a few sentences, A. E. van Vogt introduces an alien world, an alien monster, and a disturbing sense of menace. 'The Black Destroyer' was published in *Astounding Science Fiction* in July

1939. Inspired by Darwin's voyages, it was expanded into the novel *The Voyage of the Space Beagle*.

Aliens are not just literary furniture; they are usually present to make a point. It may be trite (it can be wise/foolish [delete where inapplicable] to fear the unknown) or subtle (do not assume that strangers share our attitudes and customs). It may be political (maltreatment of aliens as criticism of human colonialism or racism), or social (normal behaviour of 'filthy' aliens illuminating our puritanical tendencies, as in Brian Aldiss's *The Dark Light Years*, whose aliens wallow communally in their own excreta because it provides essential lubricants for their skin). Van Vogt's main point was to advocate holistic thinking as opposed to narrow specialism. He embodied this idea in a fictional and lightly sketched approach to knowledge, which he called nexialism. The ship's nexialist, the sole crew member proficient in this new field, is constantly discriminated against by his more narrowly specialist companions, who see his all-encompassing field of science as vague and flaky. But it is nexialism that ultimately leads to the catlike beast's defeat.

Van Vogt's story lies firmly within one of the main categories for SF about aliens: *first contact*. In these tales, humans and aliens, each blissfully unaware of the other's existence, meet. The main thrust of the story is how they handle the experience. The fun comes from devising unusual circumstances for the encounter, inventing imaginative aliens, and playing these two ingredients off against each other.

Another common category is *alien invasion*, a militaristic form of first contact. In most cases, *they* are aware of *us*, but we are not aware of them until the 10-kilometre spaceship is hovering over Washington/Berlin/Tokyo, backed up by a vast invasion fleet. The progenitor of all such stories is H. G. Wells's *The War of the Worlds*, in which Martians land their cylindrical spaceships near London. Their appearance is horrific: 'a big, greyish, rounded bulk, about the size of a bear ... [it] glistened like wet leather … There was a mouth under the eyes, the brim of which quivered and panted, and dropped saliva.' The Martians attack with tripodal fighting machines and a deadly heat-ray; a swarm of refugees heads out of the capital towards the coast. We learn that the Martians are little more than disembodied brains with 16 tentacles, which feed on fresh blood. Ultimately humans are saved by a Martian blunder: the invaders have no immunity to earthly bacteria.

Sometimes we are the invaders. A classic instance is Robert Heinlein's *Starship Troopers*, in which human armies drop from the heavens to exterminate alien races (referred to by derogatory terms like 'skinnies' and 'bugs'), with extreme violence, zero compassion, and every indication of relish. The book appears to be a propaganda vehicle for Heinlein's right-wing views, and was and still is offensive to many. However, irony is easily misread, and it has never been totally clear whether Heinlein was advocating crass militarism or exposing its immorality. The book remains controversial more than fifty years after it was published. Its aliens are cardboard caricatures, but we see them only through the eyes of the troopers who are slaughtering them.

Sometimes the alien invaders are benevolent. The classic story here is Arthur C. Clarke's *Childhood's End*. As their spaceships hover silently above Earth's major cities, the Overlords enforce peace on humanity without ever appearing in person. Only when our world has become a Utopia do they reveal themselves: they are black, with leathery wings, horns, a barbed tail – the race-memory image of devils. We are told, firmly, 'The stars are not for Man.' We are too untrustworthy to join the Overmind, the collective of galactic races. But the Overlords evolve a new breed of ethically superior human, and when they leave they take the children with them.

'Widescreen baroque' stories are set in a Universe already teeming with intelligent life forms, with an established – though often fragile – political order into which humans innocently blunder, causing mayhem. In David Brin's 'uplift' series (*Sundiver*, *Startide Rising*, *The Uplift War*, plus three later titles), humanity suddenly becomes aware that ancient races carved up the Five Galaxies a billion years ago. There is an elaborate, long-established pecking order; almost literally, because one such race, the cruel and fanatical Gubru, is avian. The sole route for new races to join the galactic club is uplift by existing members, known as patrons, through genetic modification and other technological interference. As payment, the uplifted are indentured to their patron for 100,000 years.

A fourth category is simpler: *alien as monster*. Here the role of the alien is to scare the reader and showcase the indomitable human spirit in the face of dire adversity. Or, more often, to have some mindless fun with gore and extreme violence. Aliens can play several roles simultaneously – Coeurl's is partly that of monster, but mainly to illustrate the superiority of nexialism.

Science-fictional aliens are primarily driven by narrative imperative, with occasional gestures towards scientific realism. There is nothing wrong with that; Shakespeare did the same with regard to historical realism, and it did no harm either to his reputation or his plays. Some SF authors construct an 'invisible book': an elaborate scientifically consistent setting, not presented to the reader in any detail, as deep background for their tales. Hal Clement (Harry Clement Stubbs) was famous for his meticulous planning of entire alien worlds and societies. Other writers leap in regardless and let the world unfold according to the needs of the story, with occasional blunders.

Narrative structure requires humans and aliens to *interact* to create a story. The easy solution is to invent creatures that inhabit, or at least can survive on, Earth-like worlds. More imaginative approaches have also been tried; James White's Sector General stories centre around an intergalactic hospital engineered to provide hundreds of different environments – any extreme of heat or cold, any level of gravity, any kind of atmosphere. All species have a four-letter classification: humans are DBDG, while Illensan chlorine-breathers are PVSJ. Dr Prilicla, classification SRTT, is an empath, able to sense his patient's feelings.

Most SF aliens fall into a small number of basic categories. Intelligent humanoids are common, often differing from us in relatively trivial ways such as green or blue skin, huge eyes, unusual height, excessive aggression or timidity. The Sirians in Eric Frank Russell's humorous *Wasp* are much like us, but with purple faces, pinned-back ears, and bow legs. This resemblance is crucial to the plot, allowing the minimally disguised Terran James Mowry to infiltrate and sabotage Sirian worlds. Almost as common are aliens like Coeurl, modelled on terrestrial animals but with a few exotic add-ons: cat-like, bird-like, lizard-like, insect-like. The Kzin in Larry Niven's *Ringworld* series (and other 'Known Space' tales) are modelled on tigers, and their gut reaction in tricky situations is 'scream and leap'. They develop over the series and it is fascinating to see them struggling to control their aggressive instincts.

Next in order of strangeness come aliens that differ significantly from terrestrial life forms, but still live on planets. Clement's *Mission of Gravity* sets the standard. Mesklin is a rapidly spinning, highly eccentric ellipsoid, whose distorted shape and centrifugal force cause its surface gravity at the poles to be seven hundred times that at the surface of the Earth. At the equator, however, humans can tolerate the force of gravity for a few hours, because it is 'only' three times as strong as the Earth norm. This makes direct contact between humans and aliens possible. The intelligent Mesklinites are similar to centipedes, keeping low to the ground to survive these huge forces. A galactic research project run from Earth enlists their aid to recover a gravity-probe that has crashed near the South Pole. Their adventures allow Clement to examine the physics of a high-gravity world, and a twist at the end reveals the natives to be a lot smarter than humans think they are!

Alien monsters are often drawn from human mythology, a rich and psychologically resonant source of existential terror.

Ancient cultures often depicted gods as a weird mixture of parts of different creatures, such as a human with a jackal's head, or a winged lion with a human face. Early SF writers often defaulted to similar plug-and-play alien building; Coeurl is part cat, part octopus. It is still the default in SF movies: when it comes to subtlety, Hollywood is about fifty years behind the written word (see Chapter 15).

Masters of hard SF – where the science is expected to be correct, save for the time machine, warp drive, or other revolutionary innovation that propels the story – pay a great deal of attention to the physical sciences, but tend to skimp on the biology. There is a good reason for this. You can imagine a new creature, and endow it with a very wide variety of attributes – six legs, five eyes, scales, feathers – without obviously violating any biological principles. In contrast, you can imagine a new chemical element, but you then need a PhD in quantum mechanics to figure out how it behaves. However, this flexibility is deceptive. Biology, too, has constraints, notably evolution. And here, many stories miss a trick. It seems unlikely that van Vogt wondered how his monster *evolved* the ability to suck the life force ('id') from other creatures. His aim was entertainment, not scientific plausibility. Frank Herbert never gave a satisfactory explanation of how the gigantic sandworms in *Dune* could exist on a desert planet, even though a key character was an Imperial Planetologist. Wells assumed that his Martians would be susceptible to terrestrial bacterial infection, without considering the deep evolutionary relationship between parasite and host, and failed to ask whether human blood could plausibly be a source of nourishment for creatures from another world.

Biologically realistic themes are becoming more common, however. *Legacy of Heorot*, by Larry Niven, Jerry Pournelle and Steven Barnes, is rooted in ecology. Humans colonise the fourth planet of Tau Ceti, and name it Avalon. They make their base on Camelot, a small island, to assess the local ecology, which seems benign. But when a calf is found dead with its bones neatly sheared off, they discover the presence of a predator much like a Komodo dragon, but with a thick spiked tail. Then the monster accelerates to an incredible velocity, and kills one of the colonists. They call it a grendel, after the monster in *Beowulf*. Grendels store an oxygen-rich chemical, *speed*, which powers their unbelievably rapid movement. Initially grendels are seen as mindless monsters, but in the sequel, *Beowulf's Children*, second-generation colonists, born on Avalon, start to understand them better, and the monsters become subtler. The underlying theme of the books is the need to view creatures in the context of their ecosystems. Disrupting a complex network of interactions between species can lead to unforeseen results.

The same is true on a more personal level, a theme that Philip José Farmer explored in 1952 with *The Lovers,* a controversial short story published in *Startling Stories* that examines alien reproductive biology. On the planet Ozagen, inhabited by pale green aliens, the human Hal Yarrow meets Jeanette Rastignac, an entirely human-looking woman. Contrary to the prevailing religion, the Sturch, he has a prolonged and passionate sexual relationship with her. She insists on certain precautions, but he secretly gets round these and she becomes pregnant. Only then does he discover, too late, that she is not human, but an alien mimetic parasite. When her species reproduces, larvae grow inside her body and eat her from the inside. Yarrow's grief triggers a revolt against the Sturch and an Ozagenian insurrection, neither of which assuages his conscience. The mimicry would need to operate on a deep biochemical level for the union to be fertile, but Farmer does pay some attention to the biological underpinnings.

Sex in SF was considered offensive by many in 1952, but by 1979 speculating about the sexual habits of aliens had become mainstream. John Varley's trilogy *Titan, Wizard, Demon* is set inside a giant wheel-shaped artefact orbiting Saturn, filled with an endless variety of strange creatures. Among them are titanides, like centaurs but with the twist that both the human at the front and the equine at the back possess functional genitalia. This led him to include a catalogue of combinatorial possibilities going well beyond the *Kama Sutra*, especially when it comes to group sex, with its permutations of foremothers, hindmothers, forefathers, and hindfathers.

The most imaginative aliens are plain weird. In *A Fire Upon the Deep*, by Vernor Vinge, we meet the doglike tines. They have long slender necks and ratlike heads. They operate in small packs, which often behave like a single entity. Each tine has a tympanum, a stretched membrane like a drum, which lets it translate its thoughts directly into sound waves, which it broadcasts to the rest of the pack. Here Vinge is encouraging his readers to use their imaginations, by introducing an apparently supernatural ability and then explaining it in terms of orthodox physics. He also reminds us that aliens could reasonably have abilities very different from our own.

Larry Niven's Outsiders thrive in the cold vacuum of space. Their bodies contain liquid helium, and they get their energy thermoelectrically by lying with one end in sunlight and the other in shadow. They are thought to have evolved on a small, intensely cold world; as evidence they have leased Nereid, a moon of Neptune, from the Earth government. They are galactic information sales-entities, specialising in faster-than-light interstellar transport systems. They have no business ethics and will sell anything to anyone if the price is right.

Arthur C. Clarke's 'Out of the Sun' tells of a huge jet of gas, expelled from the Sun by what today we would probably call a coronal mass ejection; Clarke describes it as 'an explosion of a million H-bombs'. The sole point of the story is the human observers' gradual realisation that the core of that vast mass of gas is, in some weird way, *alive*.

In *Dragon's Egg*, Robert L. Forward outdoes Clement with the cheela, who live on the surface of a neutron star. Such stars consist almost entirely of neutrons, and they form when a large star collapses under its own gravitational field but lacks the critical mass to become a black hole. The result is about 10 kilometres across with twice the mass of the Sun. Forward's exotic aliens provide a vehicle for exploring the physics of neutron stars, not always convincingly. For example, the storyline requires the cheela to live their lives at breakneck pace, about a million times faster than humans. As justification: the star rotates five times per second, so its 'day' is roughly half a million times shorter than ours. Forward does not explain how this rapid timescale can be consistent with the relativistic effect of the star's huge gravitational field, which slows time to a crawl at its surface. However, he does state that, in his fictional world, the topic is 'still a subject of debate among experts, since the cheela physiology is so drastically different from human physiology'. The timescale allows the cheela's abilities to overtake those of the humans observing them. The students quickly transcend their teachers; discovering five black holes inside the Sun, the cheela benevolently remove them before they devour our star, using technology beyond human comprehension.

Stephen Baxter's Xeelee sequence explores even more esoteric realms of modern physics. The Xeelee can do (very!) heavy engineering with black holes and event horizons, allowing them to manipulate time by constructing closed timelike curves. They use these as a weapon in a vast cosmic war with photino birds, who are made of dark matter and live deep inside stars. Humans, initially unaware of either protagonist, discover the existence of the Xeelee through high-tech artefacts that they have casually discarded. As humans become the second most advanced race in the Universe, a xenophobic doctrine that places survival of the human race above all other considerations leads them to wage a massive war against the Xeelee, ranging across space and time. We remodel ourselves to inhabit other regions of the multiverse, pocket Universes causally disconnected from all others and thus immune to invasion via closed timelike curves.

'How unlike the home life of our own dear queen,' as one of

Victoria's ladies-in-waiting remarked of Shakespeare's *Antony and Cleopatra*. Baxter's meticulous imagination runs well beyond anything that is scientifically likely, but it comes as a welcome antidote to those blinkered astrobiologists who think that all intelligent life must be very like us. Clarke's short story was aimed at the same point: his alien, ejected from the Sun, was powered by electricity, but 'only the pattern is important; the substance itself is of no significance'.

Blurring the boundaries between living and non-living opens up new realms of speculative fiction. In *Great Sky River* and its sequels, Gregory Benford paints a dismal future for humanity: a battle to the death against implacable aliens that are intelligent but not biological. A mechanical culture is dedicated to eliminating organic creatures from the Universe. Remnants of humanity on Snowglade have survived climate change, brought about by dust clouds that the *mechs* deliberately introduced into the planet's orbit, but now the Family must stay continually on the move to evade the malevolent machines. The setting leads to a great deal of high-adrenaline action, but also permits the examination of deeper issues about machine intelligence when one of the principal human characters has to experience what it is like to have a mechanical 'sensorium' – an artificial mind.

Series can change character as the writer develops the underlying themes. In Orson Scott Card's *Ender's Game*, Andrew ('Ender') Wiggin, led to believe he is playing an elaborate computer game, defeats the invading wasp-like Buggers and destroys their homeworld. On learning the truth, he is appalled by his act of xenocide. In the sequel, *Speaker for the Dead*, he discovers that the extermination has not been total by finding the pupa of a Bugger hive-queen. To atone for his sin he becomes a wandering Speaker for the Dead, relating the stories of beings who have died. In an act that most would consider a betrayal of the human race, he secretly carries the pupa with him, seeking a suitable world for the hive-queen to recreate the Bugger race. A series that began as a fairly standard shoot-'em-up space war moves to a far deeper emotional and ethical level.

On the surface, SF stories about alien creatures and civilisations often appear to be little more than futuristic Cowboy-and-Indian tales, albeit with hardware more interesting than Colt 45s and bows and arrows. But, as the stories described above make clear, the main role of aliens in well-crafted SF is to provide new and imaginative ways to examine what makes us human. Aliens provide problems for us to overcome, and act as a mirror in which we can examine our own faults and foibles. How we treat aliens, or react to their presence, reveals a lot about ourselves.

We have met the alien, and it is us.

LIFE AS WE KNOW IT

10

Randomness versus Complexity: The Chemistry of Life

Andrea Sella

When considering the question of whether aliens might exist beyond our planet, one of the most crucial issues to clarify is just how versatile the chemistry of life is. It's tempting to see chemistry as simply the random movement of particles, but, as we shall see, this is far from the case: in fact, chemistry tends towards complexity and order in a way which sheds light on both our own origins and the possibility of life on other planets. In what follows, I'll address this and many other questions, such as: Does life *have* to be carbon-based? Does it require water? Are there other chemical elements and compounds that can do an equally good job? Is there some magical spark that is needed to turn chemistry into biology?

In 1871 Charles Darwin mused in a letter to his friend Joseph Dalton Hooker about how life on Earth might have started in 'some warm little pool with all sort of ammonia and phosphoric salts …' His musings came at a time when there was a fierce debate over the possibility of spontaneous generation of life. On the one hand, Louis Pasteur had shown that it did not occur in a sealed flask. But some scientists argued that Pasteur's experiments were irrelevant: the timescale required for the chemicals of life to materialise was far longer than the duration of his experiments.

However life began, we have to assume – rather as Darwin and his contemporaries did – that an assortment of chemical components somehow combined to produce the 'molecules of life' and that all living organisms are nothing more than highly sophisticated chemical systems. This idea is one that remains abhorrent to many today, just as it did 150 years ago. Life, they argue, is far too complex – far too cleverly contrived – for it simply to have arisen through the processes of chance.

Perhaps this discomfort comes from the fact that the ideas of chance and randomness have actually permeated too far into our thinking about the molecular world. As schoolchildren we are taught about the particulate theory of matter, in which independent shapeless lumps of matter (particles) dance about in gases, liquids and solids. Yet to assume the molecular world simply as being 'random' is to misconstrue some of the deepest ideas that underpin chemistry as a subject.

While Darwin was sifting the evidence for evolution by natural selection, another giant of nineteenth-century science, the Austrian Ludwig Boltzmann, was wrestling with a theory to connect the general qualities of substances – such as viscosity – with the idea that, at the smallest scales, everything is ultimately composed of fundamental building blocks: atoms. He did this when many of his contemporaries thought the very idea of the existence of atoms and molecules was insane. Nevertheless, Boltzmann developed his ideas in thermodynamics, which connected concepts such as the pressure, temperature and volume of a gas with the motion and collisions of its constituent atoms or molecules. His insights had profound implications for our understanding of how chemical reactions take place. One of Boltzmann's students, the Swedish chemist Svante Arrhenius, found that the rate at which chemical reactions take place has a peculiar dependence on temperature. The relation Arrhenius uncovered suggested that for a chemical reaction to occur, the collision between two molecules required a minimum threshold energy. If the molecules were too slow, they would simply bounce apart unchanged. This meant that as you increase the temperature, an ever greater fraction of the molecules would move fast enough to overcome the threshold energy and ever greater numbers would be transformed, and so the speed of the reaction would increase.

Arrhenius's work ushered in a revolution in chemical thinking. Chemical reactions could now be thought of as taking place in a sort of 'energy landscape' with reaction pathways starting from higher ground and travelling down valleys towards the lower plains. Along the way there might be mountain passes to be crossed. These 'bumps' provide the barriers that control the overall rates of the reactions, because only a proportion of the molecules can get over the top of the pass. And it is the interplay of temperature and the heights of these barriers that determine which reactions are possible at a particular temperature, and which not.

If this discussion seems rather remote from the idea of alien life, then let me explain. What Boltzmann's and Arrhenius' ideas do is put limits on the temperature range in which useful chemistry – the kind that might lead to life – can take place. In essence, they imply a Goldilocks-type scenario: if the temperature is too low, chemistry will be very slow and some reactions will be impossible; on the other hand, if the temperature is too high, then all selectivity is lost as all the possible saddle points become accessible. Eventually, in the limit of temperature rising further, every chemical entity will be vaporised and fragmented into atoms. In other words, temperature is a crucial selector of chemistry, which sets boundaries on the realm of the chemically possible, and ultimately the biologically possible too.

The effect of temperature on reactions is just one reason why the idea that chemistry as a 'random' process is mistaken. Another is that the molecular world most definitely does not consist of shapeless particles. Far from it. Atoms join together to make molecules in accordance with their structure and the way their electrons are arranged. If, for example, we link carbon with hydrogen we get a bond that will not break apart spontaneously until the temperature reaches 300–400°C. In other words, this is a strong, stable bond. But it is also a very symmetrical link – the pair of electrons that holds the two atoms together is evenly shared between them.

If, on the other hand, oxygen is attached to carbon, a similarly strong bond results, but now the arrangement of the pair of electrons is very different. They now sit predominantly on the oxygen atom, giving rise to what is called a polarised bond, with the carbon slightly electrically positive (since the electrons are further away from it) and the oxygen slightly negative. Such polarisation, which is very common in chemistry, has many profound consequences. A molecule with a significant separation of charge becomes rather 'sticky' – bring two such molecules together and electrostatic attraction will ensure that the two will also tend to stick together. This example underlines how the vision of the molecular world as consisting of a seething anthill of randomly moving 'particles' is quite wrong: chemistry is far from random. In fact, the emergence of complex structures, and even of life itself, is the result of these delicate electronic effects, which have enormous consequences for more complex structures. Indeed, without this atomic versatility, there would be no chance of life evolving – on Earth or anywhere else in the Universe.

Chemistry drives emerging complexity

However, it is only in the controlled confines of the chemistry laboratory that one has the luxury of considering only a single chemical reaction at a time. In the real world of Darwin's little pond, there might be dozens of chemicals reacting with each other at different speeds, producing a variety of products.

In the late 1950s the Russian biochemist Boris Belouzov

stumbled across a family of apparently simple chemical mixtures that behaved in paradoxical fashion: when mixed together they changed colour, repeatedly flashing alternately yellow and colourless. Could these really be switching back and forth between two states? The suggestion that these reactions were going repeatedly from starting material to product and back again seemed absurd. Indeed, Belouzov was attacked by fellow scientists as a fraud and a fantasist. They scoffed that a chemical reaction could no more go back to the starting materials than a set of skis could 'change its mind' and go back up a hill. After all, they argued, chemical reactions follow a path through the 'energy landscape' that leads them to lower, more stable products. To do the reverse must violate the laws of thermodynamics.

Yet what Belouzov had stumbled across was not a chemical reaction but rather a chemical ecosystem, one in which molecules were formed and then consumed in ways that were modulated by chemical feedbacks. These processes gave rise to the oscillations in the concentration of chemical 'species', just as the populations of lions and wildebeest rise and fall on the Serengeti plains.

Almost at the same time the mathematician Alan Turing, unaware of Belouzov's work, theorised that, were a hypothetical system of this kind left unmixed, the interplay of the chemical reactions with the rate of diffusion (the random walk of molecules from one place to another) would give rise to patterns and structures. Research over the past seventy years has demonstrated how such systems can spontaneously generate patterns resembling those of tropical fish, of zebras, and of spotted cats. These patterns are the spontaneous outcome of chemistry itself. Along with the gradual linking and integration of chemistry into ever more complex schemes, they are phenomena that one can expect to be repeated on any of the billions of planets predicted to have the potential for life. All that is needed is the constant flow of energy from the nearby star to keep moving molecules into the energy 'highlands'.

The wonder of water

If we have a solution of a salt and we let the water evaporate, the salt will crystallise into beautiful block-like structures of its own accord. It looks like design, but it is simply the charged sodium and the chloride ions adopting the lowest energy, and therefore most stable, arrangement. For crystals made from complex molecules, predicting the structure *a priori* is a fiendishly complex task, but the final arrangement is always determined by minimising the attractions and repulsions between the different electric charges in the molecule. And if we consider more complex structures such as larger organic molecules, like fats, then we discover that ever more fascinating arrangements can develop. In fact, the energy minimisation process discussed above can result in ordered structures and neat arrangements that appear at first sight to bear an uncanny resemblance to what we see in living systems.

One might imagine that assembling such arrangements depends only on the structure of the molecules themselves. But that would be to forget the other key component that makes this self-assembly possible: water, the solvent in which the biochemistry of life on Earth takes place. It might be thought that water simply provides the medium in which these molecules move. However, water is anything but a passive matrix.

Water owes its strangeness to an almost paradoxical mix of great stability and extreme chemical promiscuity. If that sounds like a contradiction, then that is a tribute to the bizarreness of water. On the one hand, the two bonds in a molecule of H_2O, between each of the hydrogen atoms and the oxygen atom, are about as strong as single bonds can be. But they also have the additional property that the electrons in the oxygen–hydrogen bond spend most of their time close to the oxygen atom, lending it a significantly negative charge, making water exceptionally 'sticky'. One molecule links to another, leading to an endless network of strong interconnections. One unexpected consequence of the stickiness of water is that it creates structures. Let me explain. If I combine oil and water, the two don't mix. The reason is subtle. In order to mix them you would have to pull apart the network of water molecules to insert the oil molecules. In doing so the water is forced to create cages surrounding each oil molecule. It is a process that has a significant energy cost, so, very quickly, the oil molecules and the water's cluster together, like with like, and the whole system separates into the familiar two layers.

But imagine now that we construct a molecule with the shape of a typical sperm cell, with a long oily tail and a charged polar head. When this interacts with water, something astonishing happens. The polar water molecules crowd around the charged head, while the tails are forced together. These clusters have structure. They may simply be micelles – spheres with an oil-like core and the polar heads on the surface – or you may see sheets resembling the membranes of cells. If these sheets curve and close, we get cell-like entities: vesicles. In other words, it is the interplay between the structure of a molecule and the characteristics of the water that leads to the spontaneous formation of structures.

So the very nature of water is critical in driving the formation of the myriad structures we see in biological systems, from the folding of the proteins and the self-assembly of DNA helices to the segregation of cells into compartments.

The existence of water here on Earth in liquid form has been crucial for the emergence of life. But what about other planets in the solar system and their moons, such as Titan, which instead of water are known to have oceans of a different kind? Can one envisage alien life developing in other liquids? Perhaps. We can imagine oceans of liquid methane, liquid nitrogen, or even ammonia. Yet such oceans would impose very severe limitations on any developing life form. For one thing, the molecules that make up such liquids are much less 'sticky' molecules than water. The charge polarisation (stickiness) of water is so large that water melts and boils tens or hundreds of degrees above other chemicals whose molecules are of a similar size and complexity. Unless we apply pressure, nitrogen and methane boil close to $-200°C$ while somewhat stickier ammonia boils at $-40°C$. This means that they are only in liquid form when they are very cold. These cryogenic temperatures mean that any chemistry would be extremely slow. Indeed, in the laboratory, liquid nitrogen is often used to keep chemistry in suspended animation – extremely fragile molecules can be kept intact at these low temperatures; life scientists store biomolecules and even whole cells at these temperatures, precisely to hold the process of life in suspension.

The ingredients of life

We've already seen that there are aspects of chemistry which support the kind of complexity that might lead to life. But what else is needed: are there basic chemical ingredients that had to be present for the complex molecules of life to have any hope of forming?

In 1952, Miller and Urey conducted their famous experiment in which methane, carbon dioxide, water and ammonia were boiled and sparked for many weeks, resulting in a brownish solution (a sort of proto-Marmite?) from which a number of more or less complex molecules resembling amino acids and simple sugars – the building blocks of life – were isolated. Although many similar experiments had been conducted in the nineteenth century, Miller and Urey were the first to do so in the era of modern chemical analysis. Although their experiment has often been dismissed as rather naive and primitive, it nevertheless focused people's minds.

And what of carbon? To ask whether alien life would also necessarily be based on carbon is to ask the wrong question. The idea of organic and inorganic chemistry dates back to a time when we imagined that some special spark, the vital force, was needed to imbue inanimate matter with the thing we call life. In science at least, this idea of 'vitalism' has been out of favour for a long time, and the terms organic and inorganic chemistry are today as much of a hindrance to our understanding as they are a help. So carbon is not *the* element of life. It is just one of some forty elements that we know to be essential for life on Earth. Its virtue, which lifts it above all the other elements of life in terms of importance, is its versatility in forming molecules that have the necessary stability to act as robust storehouses of energy and chemical information, and within the temperature boundaries set by the liquid range of water. However, as chemistry has advanced through the twentieth and twenty-first century, we have found parallel chemistries of elements, like phosphorus and silicon.

The chemical evidence for alien life

Given that we are unlikely in the coming years to visit even the closest of places where there could be life, how would we spot it? Which group of scientists are most likely to be able to make the all-important discovery? Hunting for radio chatter is the approach adopted by SETI, yet that narrows down our options enormously to those places where life uses radios, televisions and mobile phones.

No, in our search for aliens we must turn to a set of tools that transformed chemistry from something fascinating, yet earth-bound, to a discipline that could encompass and make sense of vast tracts of the Universe. As James Lovelock and Carl Sagan proposed in the 1970s, the emergence of life on any planet would be certain to alter the composition of its atmosphere, just as it did on Earth with the sudden production of oxygen. The make-up of our atmosphere contains the telltale signature of a water-based photosynthetic network of living things.

In 1859, the same year that Darwin published *On the Origin of Species*, Robert Bunsen and Robert Kirchhoff passed the light from the Sun through a prism and realised that the dark lines in the resulting spectrum matched up exactly with the bright colours issuing from a hot flame dosed with metallic salts. Together they gave proof that the chemistry conducted on Earth held clues to the composition and behaviour of the cosmos. Forty years later, helium would be discovered on the Sun by the same method, spectroscopy, before it was isolated on Earth.

In recent months the astrophysicist Giovanna Tinetti (see Chapter 18) and her colleagues have reported the first tantalising spectroscopic evidence for the make-up of the atmosphere of an exoplanet some 70 light years away. It is a first glimpse of a world very different from anything in our solar system. Our ability to study such planets will expand hugely in the coming years as new telescopes let us peer ever further. Chemistry stands ready to help make sense of what we see.

Pliny, quoting Aristotle, wrote *Ex Africa semper aliquid novi* – 'From Africa always something new'. We can expect the same from other planets. But even if we do find evidence for alien life, we must not forget the need to nurture the life we have here on Earth, the interconnected web that supports us. The emergence of life on Earth altered the composition of the atmosphere and made possible the variety of species we see today. And yet for the past few hundred years we as a species have begun to put our own chemical imprint both on our atmosphere and the crust of the Earth. We have no idea whether our ecosystem is one of billions or whether it is unique in the Universe. But for us, it is the perfect place, the one for which we are ideally adapted. Let us not lose focus on that most precious place we call home.

11

Electric Origins in Deep-Sea Vents: How Life Got Started on Earth

Nick Lane

'I can't define it, but I know it when I see it,' said US Supreme Court Justice Potter Stewart, on the subject of hardcore pornography. He could have been speaking about life itself, which is, if anything, even more difficult to define. Could a wild fire, for example, be considered alive? Plainly not, although it satisfies some of the standard criteria, such as 'feeding', 'growth' and 'reproduction', as for that matter do growing crystals. We 'know' they are not alive, but we are hard put to come up with a strict definition that excludes them. The opposite applies to viruses. They look like tiny machines, as carefully designed to carry out their task as a lunar landing module. The way they take over the machinery of cells to turn out thousands of copies of themselves is hard not to describe as purposeful. Design, purpose: these are loaded words but they can hardly be ascribed to an inanimate force. Yet viruses lack their own metabolism – they are strictly inanimate – and so are excluded from many definitions of life.

Let's put aside even more equivocal cases such as computer programs, and simply acknowledge that life is hard to define. Does that make studying the origin of life a problem? Yes, insofar as there is no agreement about what exactly we are trying to explain, and, equally, what might constitute life on other planets. No, insofar as the very many steps *en route* to the first living cells form a continuum: there is no single point at which we can unequivocally say that a complex molecular system is all-of-a-sudden alive. The earliest steps on this continuum were certainly not alive. So what exactly were they then? They must have formed some kind of environment that was conducive to the next steps, an environment containing the potential for life itself – a 'seed' of life. From life on Earth, can we say anything about what seeds life? And if so, what might that say about the nature of aliens?

To my mind, the problem with most definitions of life is that they exclude the environment, the seed of life. Take the NASA working definition: 'a self-sustaining system capable of undergoing Darwinian evolution'. Self-sustaining? At the very least, this does not emphasise the fact that life is always sustained by its environment. Not only did the first steps towards life depend on a facilitating environment, but even today we cannot cut our umbilical cord to the environment that sustains us. Like all living organisms, from bacteria to plants and animals, we need to respire continuously, just to go on living. The deep fear of drowning or asphyxiation is the fear of being separated from our environment for more than a few seconds. While a handful of living things have succeeded in separating the two for a while, in forming metabolically inactive spores that are capable of returning to life when conditions are suitable, they cannot do so indefinitely; and they are just as inseparable from their environment when they do return to living. Living is the key word here. Life is for living, and in thinking about the origin of life, the origin of living – actively exploiting the environment to power growth – is a far more meaningful term.

All life exploits its environment to make copies of itself. I don't pretend that's a definition, but it is a fruitful way of seeing the world, and it is satisfying in that it embraces viruses as well as living cells. How so? Viruses exploit an extremely rich local environment, the insides of a cell, awash with all the energy and machinery they need to replicate themselves. They can afford to pare themselves down to a bare minimum because their environment does all the work. At the other end of the spectrum plants also exploit their environment, albeit for very little: they need sunlight, water and carbon dioxide, and not much else. They have to be extremely complex in their biochemistry to furnish themselves with everything they need to grow in so sparse an environment. As a rule of thumb, the less the dependency on the environment, the greater the biochemical complexity of the organism must be. But it is still dependency: deprive plants of light or water and they will die as surely as you or I when deprived of oxygen. Like viruses, we are all parasites on our environment, and ultimately on our dynamic 'living' planet.

So how exactly does life live? There seem to be as many ways of living as there are living things, but down at the level of the basic circuitry of cells, that is not true at all. In fact it is shocking that all life on Earth shares exactly the same mechanism of harnessing energy from the environment and using it to power growth and reproduction. The chemical reactivity of the environment is used to charge up life's batteries, giving cells an electrical charge across thin membranes. The distance this charge operates across is so short (5 millionths of a millimetre) that if you were to shrink yourself down to the size of a molecule, the intensity of the electrical field you would experience would be about 30 million volts per metre, the same as a bolt of lightning. It might seem quirky, or even reminiscent of Frankenstein's monster, but this electrical charge on biological membranes is as central a feature of life on Earth as DNA or the genetic code itself. Unlike DNA, though, this pervasive charge points to a specific environment in which life may have emerged on Earth, an environment that might be found on another 40 billion planets in the Milky Way alone.

May the proton-motive force be with you

The idea that all cells are powered by a form of electricity was one of the most revolutionary of twentieth-century science. It was pioneered over several decades from the early 1960s by the eccentric English biochemist Peter Mitchell, who upset his contemporaries so much that the field degenerated into a rancorous conflict known as the Ox Phos wars (from 'oxidative phosphorylation', the mechanism of respiration). Mitchell was eventually awarded the Nobel Prize in 1978, and his discovery was hailed as 'the most counter-intuitive idea in biology since Darwin, and the only one to compare with those of Einstein, Heisenberg and Schrödinger'. Yet in essence Mitchell's ideas were simple, and were rooted in an almost naive question about the difference between inside and outside.

Specifically, Mitchell wondered, how do bacteria keep their insides different to the outside world? He realised that they do so by actively pumping molecules in or out of the cell, across the external membrane. Active pumping costs energy, and is selective: specific molecules are recognised and carried across the membrane, in much the same way that selected passengers pay to be ferried across a river. Mitchell's genius was to see that the same basic principle applied not only to bacteria but also to respiration, which was known to require a membrane, although the reasons for that were shrouded in mystery. Just as it costs energy to actively pump anything out of the cell (producing a difference between the inside and outside), Mitchell realised, energy is released if it is allowed to flow back in again, dissipating that difference. The energy released could be harnessed to power work.

That's how respiration works. In this case, protons are actively pumped out across the membrane. Protons, you may recall, are the positively charged nuclei of hydrogen atoms, given the symbol H^+. Pumping them out of a cell produces a difference in proton concentration between the inside and outside, but also an electrical charge across the membrane, as protons carry a positive charge. Push them outside, and the outside becomes positively charged relative to the inside. And that is where the pervasive electrical charge on our membranes comes from. Once protons have been pumped outside, they want nothing better than to get back inside again, to collapse the difference in charge and concentration. Mitchell called this the proton-motive force. It is the most fundamental force in life. We pump an incomprehensible 10 billion trillion protons every second across the membranes in our mitochondria, the powerhouses of our own cells – nearly the same number as there are stars in the known Universe. Gram per gram, bacteria pump even more. Every living being on this planet is powered by the same electrical force field, the proton-motive force. It operates continuously, second by second, life by life, passing on the living flame from generation to generation, a never-ceasing flux of protons from the first stirrings of life on Earth 4 billion years ago.

So where does the energy come from to pump all these protons, to maintain the force through all of these lives? In our own case, from burning food in oxygen: respiration. We strip electrons from food, and pass them along within that same membrane, hopping from one carrier to the next, until they are ultimately brought together with oxygen, breathed in deeply to each of our cells for that reason alone. The flow of electrons to oxygen – another electrical current – is what powers the extrusion of protons across the membrane. Stop breathing and the flow of electrons ceases: now nothing powers the pumping of protons, the force collapses, and that's that. The end. Collapse of the proton-motive force is the best definition of death. The unbroken flow of electrons and protons that connected the origins of life with our own lives peters out with our demise.

Most bacteria don't need oxygen or even food to generate a proton-motive force – they can use other gases or even rocks to power up their force field. We can study genes to trace back the origins of the force field to the Last Universal Common Ancestor of all life, fondly known as LUCA. LUCA is not directly relevant to the origin of life, because she was already a cell with genes and proteins, and so quite complex – undoubtedly alive. But still, LUCA was very early in the history of life, and if she was already powered by the proton-motive force, then the force must have arisen earlier still, perhaps in those shadowlands between non-living and living. The sheer complexity of modern respiration howls in opposition to this notion, and that is why the idea has received little attention. But the more that we learn about the curious make-up of LUCA, the more she points to a primal electrical force – and not just here on Earth, but right across the Universe.

Looking for LUCA

You may have heard about the three great domains of life: eukaryotes (all organisms with large complex cells, including plants, animals and fungi), bacteria, and a third group called archaea, which look the same as bacteria but are strikingly different in their genetics and biochemistry. We now know that eukaryotes – our own type of cells – are in fact cobbled together from bacteria and archaea, the result of a freak interaction between these two types of cell. Fascinating as that is, it's totally irrelevant to our story here: eukaryotes arose 2 billion years later and have nothing to do with the origin of life. In the last few years it has become abundantly clear that there are only two primary domains of life, the bacteria and the archaea. Both are genetically diverse groups comprising tiny single-celled organisms, fabulous in their biochemical sophistication but stunted in their morphological complexity.

If you're not familiar with bacteria and archaea, they might sound misleadingly dull and primitive. They're anything but. They dominated the first 3 billion years of life on Earth, and between them they invented all the most fundamental biochemistry, from photosynthesis and nitrogen fixation to respiration. Even today we couldn't live without them. But most importantly to us now, we can compare the detailed biochemistry of the two groups to try to work out what their common ancestor – LUCA – might have looked like. For example, bacteria and archaea share the use of DNA (deoxyribonucleic acid) as the hereditary material, which encodes the sequence of building blocks (amino acids) in proteins. The code is exactly the same in both groups – it's known as the 'universal genetic code'. We can infer that LUCA already had the genetic code, DNA and proteins.

So what else did LUCA have? Well, we can be fairly sure that she was an autotroph, which means that she got everything she needed to grow from inorganic matter in rocks and gases, not by 'eating' organic material. In particular, the most ancient bacteria and archaea seem to have powered their growth with the gases hydrogen and carbon dioxide – probably that's what LUCA did, too. We can be more or less certain that LUCA was not photosynthetic (she did not depend on the Sun) as that sophisticated skill is found only in bacteria, and not archaea. But hydrogen and carbon dioxide do not react very easily. LUCA seems to have used the proton-motive force – the electrical charge on her membranes – to make them react together, with the help of mineral structures composed of iron and sulphur that worked as catalysts, speeding up natural reactions. These traits are found in both groups, and so were probably present in their common ancestor, LUCA, too.

But perhaps the single most interesting fact about the two domains is how different they are to each other in several other respects. LUCA used the electrical charge on her cell membrane to drive the slow reaction between carbon dioxide and hydrogen – yet this membrane, so central and important, is confoundingly different in bacteria or archaea today, as is the machinery that generates the charge.

How can we make sense of this? There are several conceivable ways, and little agreement among those working on the problem.

But one possible explanation is compelling because it points straight to the origin of life in one very particular environment. LUCA could depend on an electrically charged membrane, while at the same time lacking the machinery to generate the charge, if that charge was provided for free by the environment.

Electric origins

Just such an environment, a particular type of hydrothermal vent, was first proposed in the late 1980s by Mike Russell, now at the NASA Jet Propulsion Laboratory in Pasadena. No matter that such vents were unknown at the time, at least in the deep oceans – a decade later, in the year 2000, a new vent field was discovered close to the Mid-Atlantic Ridge, which corresponded to all Russell's predicted criteria. Named 'Lost City', this vent field was the product, not of volcanic activity, but of a chemical reaction between rocks in the oceanic crust and seawater. This reaction gives rise to strongly alkaline hydrothermal fluids, equivalent to kitchen bleach, bubbling with hydrogen gas. That might sound toxic, but the hydrogen gas is just what the most ancient bacteria and archaea need to grow. Even more excitingly, the vents at Lost City are riddled with great labyrinths of tiny pores, bounded by thin inorganic walls. The pores are not only cell-like in their structure, they even have an electrical charge across their bounding walls – a natural proton-motive force, produced by the difference in proton concentration between the hydrothermal fluids (alkaline technically means 'proton-poor') compared with the relatively acidic (proton-rich) seawater, which mix together within the vents.

The vent pores are precisely analogous in their structure to the most ancient bacteria and archaea that, intriguingly, still live in these primordial environments today. The only difference is that today the presence of oxygen prevents the kind of chemistry that might once have got life going. But 4 billion years ago, before photosynthesis began producing oxygen, could a natural proton-motive force have driven life into existence? The idea is appealing in the context of decades of work on the origins of life. That work has been driven by pragmatism – the chemistry that works well in the lab. Start with high-energy molecules such as cyanide, excite them into reacting by bombarding them with ultraviolet radiation, and it is possible to synthesise most of the basic building blocks of life. It works well – but after decades of research, the picture emerging is disturbingly dissimilar to anything we know from life itself. No known form of life on Earth uses cyanide as a source of either carbon or nitrogen, and none is powered by ultraviolet radiation. The reaction pathways that have been painstakingly pieced together by prebiotic chemists look nothing like the biochemical pathways found in actual cells. And getting from a dilute soup of these molecules to cell-like structures that grow and divide has proved a bridge too far.

Yet the opposing scenario – starting with the molecules and biochemical pathways used by living cells – has not worked well in the lab either. Carbon dioxide and hydrogen remain stubbornly unreactive, for all that life uses them as the basis for everything. What has not, until recently, been simulated in the lab is the structure of cells themselves – the electrical charge on membranes, the proton-motive force. How might it work? The best clues surely lie in the cells themselves. Guided by the pioneering work of Bill Martin at the University of Düsseldorf on the detailed metabolism of the most ancient cells, several groups around the world (including my own at University College London) are developing reactors to test the idea that a natural proton-motive force in hydrothermal vents could have driven the beginnings of life. We are getting some positive results, the first tantalising hints that natural proton gradients really could drive the reaction of hydrogen with carbon dioxide to form simple organic molecules. But, exciting as it is, that is not the point.

The point is that these ideas link a living planet with living cells. I said that this type of vent is produced by a chemical reaction between rock and water – a reactive environment, and possibly one of the most ubiquitous environments in the Universe. The rock in question, olivine, is among the most abundant minerals in interstellar dust, and makes up the greater part of the Earth's mantle. Water, too, is everywhere. Put them together, on any wet rocky planet, and they will react on a planetary scale. In our own solar system there are signs of these reactions on Mars (despite it having lost most of its water), as well as on the icy moons of the gas giants: Enceladus, Titan and Europa. Carbon dioxide, too, is abundant in the atmosphere of most planets in the solar system. This 'shopping list' for life – rock, water and carbon dioxide – is about as short and non-specific as could be. Yet these conditions should produce the right kind of hydrothermal vents, with natural proton gradients across thin inorganic barriers, driving the reaction of hydrogen and carbon dioxide to make organic matter in cell-like pores on any wet rocky planet.

Earth is not unusual. On those 40 billion wet rocky planets in the Milky Way, hydrogen is bubbling from the ground and reacting with carbon dioxide, powered by the unceasing flow of electrons and protons, elementary particles of the Universe. The same forces are alive there. And so we will know life when we see it, for the aliens will be electric too.

12

Quantum Leap: Could Quantum Mechanics Hold the Secret of (Alien) Life?

Johnjoe McFadden

How hard is life?

Life is complicated. There are about 9 million known species on Earth, but vastly more unknown species; each a highly complex product of more than 3 billion years of evolution. However, the biosphere wasn't always so complex. At some point in the Earth's early history there was no life, only non-living chemistry. As discussed in the previous chapter, to understand that transition from chemistry to biology, we need to agree on a definition that separates the non-living chemistry from life. Unfortunately, none exists that all biologists and chemists can all agree on. The tricky question of a definition of life has been adequately explored by Nick Lane so I won't labour the point here; but, from the perspective of exploring the possibility of life independently originating on other planets, it is useful to take as our definition organisms that can replicate in non-living environments, the only kind available in any initially sterile world.

We can now use this self-replicator (in a non-living environment) definition to ask the question – how simple can life be?

The answer, as far as we know, can be found today in organisms known as mycoplasmas – tiny bacteria that are a common cause of infections in humans, such as pneumonia. Still, these are far from simple life forms. Even the most basic mycoplasmas have sufficient genes to provide the instructions to make nearly 500 proteins. Mycoplasmas are irregularly shaped single-celled organisms that are less than a thousandth of a millimetre in diameter, yet they are packed with millions of molecules, of protein, fats and carbohydrates, all linked by highly intricate regulatory networks.

The reason that living cells are complicated is because the process of self-replication is itself complicated. We know that self-replication is difficult even for machines that have been specifically built to replicate things. Modern technology has provided us with lots of devices that can replicate stuff, from photocopying machines to electronic computers to 3D printers. But can any of these make a copy of *itself*? Probably the closest to being able to do this are 3D printers. These devices can now print their own components, which can then be assembled to make other 3D printers. But, not all their components can be made in this way and they still need some help with assembling all the parts together. At the time of writing we are a long way from building a truly self-replicating machine. Indeed, within our technologically advanced world of supercomputers, space rockets and smartphones, it is humbling to realise that the only way we can construct a self-replicating entity is via that same messy business that has been going on for billions of years. The only self-replicating products of all human endeavour are … our children.

The apparent difficulty of self-replication represents a dilemma for those attempting to account for life's origin. The astronomer Fred Hoyle famously illustrated the problem by considering the probability of assembling a structure like a bacterium from the random thermodynamic processes available on the early Earth and likening its chances to that of a tornado blowing through a junkyard spontaneously assembling a jumbo jet. The junkyard could be on Earth or some distant planet; but the problem remains. It would seem that, despite its vastness, the Universe is just not big enough to generate complex cellular life by chance alone. And yet it did, at least once.

The primordial soup

A potential way out of this dilemma is to invoke the anthropic principle, which aims to answer the question of why everything in our Universe, from the laws of physics and chemistry to the unique conditions here on Earth, appear to be so finely tuned to conspire to allow us to be here, contemplating our remarkable existence. Often, some form of parallel Universes theory is invoked: this is the idea that there is a near-infinite number of different Universes with varying conditions, most of which are not suitable for life. However, the one in which we find ourselves had to be suitable – or we wouldn't be around to contemplate it. However, even in a life-compatible Universe like our own, life may be hard to get going. But as Paul Davies later argues more carefully in Chapter 13, if it happened once then why not again, and again, elsewhere in the Universe? If we live in such a '*life is easy*' Universe then our experience on Earth is likely to have been mirrored by similar emergences of life on alien planets. We could view this as the alien-optimistic scenario.

The second option is that the laws of physics and chemistry and the values of the fundamental constants, like the strength of gravity or the electric charge of the electron, are necessary but not sufficient conditions for life to appear. Maybe life required a further, terrestrial, throw of the dice that generated an extraordinarily rare chemical configuration that sparked the emergence of life uniquely here on Earth, as described by Matthew Cobb in Chapter 14. In this '*life is hard*' scenario, we are likely to be alone in the Universe.

So, did life require two lucky throws of the anthropic dice or just one? Is life easy or hard? As Matthew Cobb points out, the fact that all living organisms have descended from a single ancestor suggests that life on Earth had a single unique origin. But knowing of just one example of the appearance of life on Earth doesn't preclude the possibility that there were a variety of preceding successful attempts. There is only one species of the genus *Homo* today (*Homo sapiens*); but there were many others in the past, all of which became extinct. Likewise, and further back in time, alternative solutions to the origin of life may have been similarly outcompeted by our successful ancestors.

A possible argument in favour of the optimistic scenario is that life emerged on Earth very soon after it became possible. When the Earth first formed about 4.6 billion years ago it was too hot to have liquid water (which both Chris McKay in Chapter 6 and Andrea Sella in Chapter 10 have argued was such an important precondition for life on our planet and potentially elsewhere too). Therefore, it was not capable of supporting life until about 3.8 billion years ago. Yet there is chemical evidence for life in the earliest rocks that were formed around this time. If life is hard and requires an extraordinarily improbable throw of the chemical dice, then it may have taken many millions or even billions of years before the right chemical configuration came together to spark life's emergence. But life did emerge, and relatively quickly, which suggests that once the conditions for life – such as liquid water – are fulfilled then life is not only possible, but probable. If that is the case then surely we inhabit a *life is easy* Universe and that if similar conditions to those of the early Earth have been generated on alien worlds – and it seems likely that they have – then life would have similarly emerged on those planets just as quickly. Aliens should be everywhere.

But to ensure that life really is easy we have to overcome Hoyle's junkyard problem. How do we randomly assemble the chemical junk to make an extraordinarily complex, and highly improbable, self-replicator?

Several biomolecules have been proposed as the very first self-replicators that could have been formed in the primordial soup. However, even the very simplest of these would still be highly complex structures. It has been estimated that the chances of generating a self-replicating molecule by random processes alone are so exceedingly small that they can be discounted. We thereby come to the crux of the origin of life problem. It isn't that it is difficult to form the chemical precursors for life, or to identify certain biomolecules capable of performing some of the necessary steps of self-replication. The problem is that any such self-replicator is likely to be only one of a handful of structures among a mind-numbingly vast number of possible ones. This is referred to as a 'search problem': how can the correct structure be found … by accident? The problem is that random searches (essentially via atoms and simple molecules bumping into each other, interacting and sticking together following the laws of thermodynamics and chemistry) are far too inefficient for a self-replicator to be made in any feasible period of time, even over hundreds of millions or billions of years.

One way of seeing this more clearly is to explore the question within a computer, where those messy and hard-to-make chemicals can be replaced by the simple building blocks of the digital world: namely, the bits that can only have a value of either 1 or 0 (or true/false or yes/no). A 'byte' of data, consisting of eight bits, represents a single character of text in a computer code and can be roughly equated with the unit of genetic code. We can now ask the question: among all the possible strings of bytes, how common are those that can replicate themselves in a computer?

Here we have a huge advantage, because self-replicating strings of bytes are actually quite common: we know them as computer viruses. These are relatively short computer programs that can infect our computers by persuading their Central Processing Unit (CPU) to make loads of copies. These computer viruses then hop into our emails to infect the computers of our friends and colleagues. So if we consider the computer memory as a kind of digital primordial soup, then computer viruses can be considered to be the digital equivalent of primordial self-replicators.

One of the simplest computer viruses, Tinba, is only 20 kilobytes long: very short compared with most computer programs. But while 20 kilobytes may be very short for a computer code, it nonetheless comprises a relatively long string of digital information since, with 8 bits in a byte, it corresponds to 160,000 bits of information. So to have any chance of generating Tinba from the random assembly of bits we would need to make at least 2^{160000} (1 followed by 36 thousand zeros). This is a mind-bogglingly large number – far, far larger than the number of particles in the Universe and tells us that Tinba could not have arisen by chance alone.

Perhaps there are very many possible self-replicating codes simpler than Tinba and that might arise by chance. But if that were the case, then surely a computer virus would, by now, have arisen spontaneously from all the zillions of gigabytes of computer code that are flowing through the internet every second. These codes are all potentially functional in terms of instructing our CPUs to perform basic operations, such as to copy or to delete – they could all potentially be versions of Tinba – yet all of the computer viruses that have ever infected anyone's computer show the unmistakable signature of human design. As far as we know, the vast stream of digital information that flows around the world every day has never spontaneously generated a computer virus.

The primordial quantum soup

One speculative idea that, if correct, might explain *how* the first self-replicator might have emerged so quickly relies on one of the strangest and yet most powerful theories in science: quantum mechanics. Well, it might. To see how, we first need to explore (briefly, I promise) some of the ideas in this theory, which is normally only used to describe the behaviour of the atomic and subatomic world. The interested reader is referred to any of a host of excellent popular science books that will provide more depth and detail (including a recent exploration of the role of quantum mechanics in biology by Jim Al-Khalili and me). Quantum mechanics is famously weird. It allows particles to be in two or more states at the same time, a phenomenon known as *superposition*; it allows them to be spookily connected or *entangled* with distant partners; it allows them to pass effortlessly through impenetrable barriers by a process known as *quantum tunnelling*. Intense efforts are ongoing to harness these extraordinary capabilities to construct new kinds of technologies such as quantum computers.

The essence of a quantum computer is that it allows a small number of particles to solve very hard problems. The quantum equivalent of a conventional computing bit is the quantum bit, or qubit, which makes use of this idea of quantum superposition of a particle being in two positions at once, or having two precise energies at once, or even doing the equivalent of spinning in both directions at the same time. Stringing together such qubits in a quantum computer means that it is turned into the ultimate parallel processor, able to be in all arrangements of 0's and 1's at the same time.

So what has this got to do with biology and the first replicator? The currently favoured solution to the origin of life problem is what is called the RNA world hypothesis. In this scenario, living cells were descended from an earlier, simpler phase of chemical evolution. So the first self-replicator wasn't a cell: it was a self-replicating chemical molecule. And it may have looked like a more basic version of a biomolecule we know today to play a crucial role within living cells, called RNA. Although no one has yet synthesised such a self-replicating molecule, most are relatively simple structures consisting of a string of a hundred or so chemical units, called bases.

So let us imagine that there exists a unique 100-base molecule that has the extraordinary property of being about to replicate itself in a feasible primordial soup such as would have been available here on Earth 4 billion years ago – and by 'feasible' I mean 'containing the necessary chemical ingredients'. The problem that life's origin now has to overcome is much simpler than making a self-replicating cell: it only has to construct the proto-replicator molecule and natural selection can thereafter kick in to evolve more complex life.

Because a molecule of RNA is a much simpler structure than an entire cell, it is easier to attempt to look at the problem mathematically. Each RNA string is made of four different bases that may be strung together in any combination. And if each of the four bases can be in any one of the 100 positions on the molecule, this means there would be 4^{100} possible structures. This is an impossibly huge number: like a 1 with *sixty* zeros after it; so any feasible primordial pond is only likely to possess a very tiny fraction of all possible RNA sequences and is therefore very unlikely to include that special proto-self-replicator. Life is still hard to make.

To see how quantum mechanics might help, we can imagine each possible molecular string of bases as a sequence or 0's and 1's or as a string of coins each of which may have landed on either heads or tails. Our proto–self-replicator is represented by a unique sequence of heads and tails along that string. Now let's say that our primordial string is constructed from qubits rather than bits. This is not as difficult as it may sound because the coding capacity of such molecules resides in a particular type of chemical bond, called a hydrogen bond, which is essentially a proton that links two atoms together. As the physicist Per-Olov Löwdin pointed out more than fifty years ago, the genetic code of DNA or RNA thereby represents a quantum code of proton positions. And crucially, as quantum particles, protons can tunnel (remember this is one of the strange quantum properties that allows particles to penetrate classically impenetrable barriers) from one coding position (0, or heads) to another (1, or tails).

Can we apply this scenario to the origin of life problem? Let's imagine the proto-genetic material as a string of qubits, rather than bits. This potentially transforms the hard chemical search problem into something that could be solved by quantum computing. Remember that a single molecular qubit string could exist as a quantum superposition of all possible configurations. A very small component of this huge quantum superposition will represent the special arrangement of these self-replicators. Thus even a tiny primordial pool would contain the self-replicator, so long as it is a quantum pool.

Of course this quantum state is pretty delicate and will quickly reduce to just one particular string configuration, which will almost certainly have the wrong molecular arrangement for self-replication. So, on the face of it, this doesn't seem to offer any advantage over classical 'making and breaking' of the molecular structure. The crucial point is that trying out a different configuration without quantum help would have to involve the very slow process of dismantling and rearranging molecular bonds. But after this collapse of the quantum state of the molecule, each of its protons will, almost instantaneously, be ready to tunnel again into a superposition of both positions to re-establish the original quantum superposition of all possible coding structures. The quantum proto-replicator molecule could repeat its search for self-replication in the quantum world continuously and extremely quickly.

So, as long as the system can slip back into the quantum world, then the making and breaking of the quantum superposition state is a reversible process, and one that is far more rapid than the classical making and breaking of chemical bonds.

And there is one event that will terminate the quantum coin tosses. If the quantum proto-replicator molecule eventually collapses into that very special and rare self-replicator state, it will start to make copies of itself and this will force the system to make an irreversible transition into the classical world. This is a subtle point – one that the founding father of quantum mechanics, Niels Bohr, referred to as an 'irreversible act of amplification'. By replicating itself, the quantum coins will have been irreversibly thrown and the first self-replicator will have been born into the classical world.

Quantum mechanics may thereby allow the very hard search for a proto–self-replicator to be much more efficient than it could be in a classical chemical search scenario. For the quantum scenario to work, the primordial biomolecule, our proto–self-replicator, must have been capable of exploring lots of different structures by quantum tunnelling of its particles into different positions. Do we know of molecules that are capable of such a trick? Yes, we do: the electrons and protons in the biomolecules of living cells that are still held relatively loosely and can thereby tunnel into different positions. Indeed, as I have already mentioned, the protons in DNA and RNA are also capable of tunnelling. So we might imagine our primordial self-replicator to be something like an RNA molecule that was loosely held together by hydrogen bonds and weak electronic bonds that allowed its particles to travel freely through its structure to form a superposition of its trillions of possible configurations.

There are, of course, many difficulties with this scenario but, as explained above, there are many difficulties with all accounts of the origin of life. One merit of the quantum proposal is that it could account for how life managed to emerge so quickly here on Earth. And if quantum mechanics did indeed help to solve the problem of finding the first self-replicator on Earth, then there's no reason to believe it wouldn't have played the same role on other planets. Of course, the conditions would have been different on an alien world, with a different atmosphere, different chemistry in its oceans, different physical cycles, etc. But, as already noted, self-replication is a general problem and unique solutions may have been required on different worlds that were suited to the conditions and resources found there. Yet the ability of quantum mechanics to explore multiple solutions simultaneously would have allowed it to compute the answer to the problem of how to make a self-replicator on any world. Alien life, with one foot in the quantum world, may indeed be common in our Universe.

13

A Cosmic Imperative: How Easy Is It for Life to Get Started?

Paul C.W. Davies

In the past decade astronomers have discovered many extrasolar planets; some estimates suggest that there could be a billion Earth-like planets in the Milky Way Galaxy alone. This has encouraged the popular belief that, with all this habitable real estate, life must be widespread in the Universe, while intelligent life, though rarer, should not be uncommon. However, just because a planet is habitable does not mean it is inhabited. A planet only becomes inhabited if it spawns life. For life to get going on an Earth-like planet all the necessary physical and chemical steps have to happen. But because we don't know what those steps are, we simply can't say how many habitable planets do, in fact, host some form of life: one cannot estimate the odds of an unknown process. Nevertheless, this state of ignorance has not stopped many distinguished scientists from professing their belief that life does, in fact, form readily under Earth-like conditions. What is their justification for this optimism?

First some history. Fifty years ago opinions were very different from today. At that time, most biologists thought the origin of life on Earth was a freak chemical accident, involving a sequence of events so low in probability that it would be unlikely to be

repeated anywhere else in the observable Universe. In his celebrated 1972 book *Chance and Necessity*, the Nobel Prize winning
biologist Jacques Monod summarised the mood by declaring that
'the Universe is not pregnant with life', and therefore: 'Man at
last knows he is alone in the Universe.' George Simpson, one of
the great neo-Darwinists of the post-war years, dismissed SETI,
the search for intelligent life beyond Earth, as 'a gamble at the
most adverse odds with history'. Monod and Simpson based
their pessimistic conclusions on the fact that the machinery of
life is so stupendously complex in so many specific ways that it
seems inconceivable it would happen more than once as a result
of chance chemical reactions. In 1981 Francis Crick, co-discoverer
of the double helix structure of DNA, echoed the drift when he
wrote in his book *Life Itself: Its Origin and Nature*: 'The origin of
life appears ... to be almost a miracle, so many are the conditions
which would have had to be satisfied to get it going.' In effect,
back in the 1960s and 70s, to profess belief in extraterrestrial life
of any sort, let alone intelligent life, was tantamount to scientific
suicide. One might as well have expressed a belief in fairies. Yet by
the 1990s sentiment had swung in the opposite direction. By then,
another Nobel Prize winning biologist, Christian de Duve, was so
convinced life would pop up wherever it had a chance he called it 'a
cosmic imperative'. And that has become the prevailing view: life
is a cosmic imperative, and the Universe is teeming with it.

Two popular current arguments that life must be common are:
(1) The Universe is so big there must be life out there somewhere,
and (2) Life must form readily because it started on Earth so
quickly. Both arguments are bogus. Let me examine them in turn.

The up-it-pops fallacy

Carl Sagan, long-time champion of the search for extraterrestrial
life, once wrote: 'the origin of life must be a highly probable affair;
as soon as conditions permit, up it pops!' And it is undeniably

true that the rapid appearance of life on Earth is entirely consistent with the hypothesis that life forms readily in Earth-like conditions: a 'high-probability genesis'. Less obvious, but certainly
correct, is that its rapid appearance is equally consistent with its
origin being exceedingly improbable, as was pointed out by the
cosmologist Brandon Carter over three decades ago. The essence
of Carter's argument is that any given Earth-like planet will have
a finite 'habitability window of opportunity' due to the lifetime
of its parent star, during which life has a chance to emerge and
evolve to the level of intelligence. On Earth, this window extends
for roughly 5 billion years – from about 3.8 billion years ago
(when the heavy bombardment of our planet by asteroids began
to abate) to about 800 million years hence (when the Sun will be
so hot that Earth will become a sterile furnace). So unless life had
got going pretty quickly, intelligent life (e.g., humans) would not
have evolved before Earth's habitability window closed, and so we
would not be here to discuss the matter. But whether or not this
means that it was likely that it did so is a different matter.

Carter's argument goes like this: when you think about it, it's
a quite remarkable coincidence that the time it has taken for life
to emerge on Earth and evolve into 'intelligent observers' (us) is
roughly the same (a few billion years) as the time that our Sun
has offered as a habitability window. There is no reason why this
should be so, since the two timescales (for evolution of complex life
on the one hand and for the interplay between nuclear fusion and
the force of gravity in a star that defines its lifetime on the other)
are entirely unconnected. This 'coincidence' suggests two possibilities: either the average time for complex life to emerge elsewhere
in the Universe is typically much shorter than the lifetime of a star,
in which case intelligent alien life should be common, or it is much
longer, in which case we got lucky and alien life elsewhere is rare.

Carter uses careful reasoning based on the subtle idea that we
have to take into account the fact that we only have a statistical
sample of one from which to draw any conclusions (since life on

Earth is the only intelligent life we know of). He argues, firstly, that life on Earth got going pretty quickly once the habitable window opened. Thereafter, a number of other very unlikely steps were needed before intelligence was attained (examples are sex, multicellularity, and the evolution of a central nervous system, all of which seem to be rare flukes). It is therefore highly improbable that intelligent life would have evolved on Earth any sooner than it has, given that there don't seem to be any steps that could have been taken much more quickly to cut down the time from the 3.5 or so billion years it has taken to go from the first biomolecule to us.

The upshot of this reasoning, which is somewhat counter-intuitive, is that the rapid appearance of life on Earth is statistically completely compatible with the transition from non-life to life being highly unlikely, a fluke that, all else being equal, would on average require a time far in excess of the duration of the habitability window to happen. In that case life elsewhere in the Universe will be exceedingly rare. Thus we can say with confidence that the rapid appearance of life on Earth is consistent both with life being easy to get going, or extremely hard. From a sample size of one it is impossible to discriminate between the two.

The big Universe argument

During a recent television interview, the renowned cosmologist Stephen Hawking said: 'To my mathematical brain, the numbers alone make thinking about aliens perfectly rational.' Hawking was articulating the popular argument that the sheer vastness of the Universe all but guarantees we are not alone. As stated, the argument is not wrong, but needs nuancing. Obviously, in an infinite (and homogeneous) Universe, the existence of a possible event, however improbable, is certain. (It has to happen somewhere; in fact, it has to happen in an infinite number of places). That applies not just to a second sample of life but to even less likely events, such as duplicates of this author, duplicates of Shakespeare and his plays, exact duplicates of planet Earth complete with its entire population, and so on. What is of interest is not the certainty of life out there somewhere, but its density. How likely is it, for example, that there is at least one other planet with life among the 400 billion stars of the Milky Way Galaxy? That's what we would like to know.

Because 400 billion is a dazzlingly large number, it makes the odds of life arising in that milieu seem high. But a moment's thought reveals how poor human intuition is on this matter. Suppose life's origin needs a specific sequence of ten rather critical and rather precise chemical reactions (surely an underestimate), and each one has a probability of occurring in the habitability window of, say, one in a hundred, then the combined probability for all ten steps to occur is one in 10^{20}, or a hundred billion billion to one against. In that case the odds of a second planet with life in the Milky Way would be negligible.

My objections to the 'up it pops' and 'big Universe' theories are both based on the assumption that the footsteps from non-life to life are random processes with well-defined probabilities attached. Perhaps this is not correct. Maybe the molecular dice were loaded by some sort of law or life principle that favours the formation of the very molecular structures life needs? The notion that nature is somehow rigged for life, or that life is 'built into' the laws of physics and chemistry – a notion sometimes called biological determinism – is a popular, albeit vague, belief, and is the basis for de Duve's 'cosmic imperative'. What can be said in its favour? There is nothing obvious in the laws of physics that singles out 'life' as a favoured state or destination. The laws of physics (and chemistry) are 'life blind' – they are universal laws that care nothing for biological states of matter specifically, as opposed to non-biological states. That doesn't mean that no 'life principle' exists in nature, but if one does it has yet to be eluci-dated. Perhaps such a principle lurks in the realm of complexity theory or information theory or in the properties of self-organising systems, but so far there is no hard evidence for it.

How, then, can we test the undeniably appealing idea of the cosmic imperative? One obvious possibility is observation. If life's origin is probable, and life is widespread, we should be able to find a second sample of it. But where? One popular suggestion is the planet Mars, which is, or once was, moderately 'Earth-like'. Unfortunately there is a complication. Because Earth and Mars trade rocks ejected into space by asteroid and comet impacts, they can also trade microorganisms ensconced within the rocks, so the two planets are not quarantined. In any case, a Mars sample return mission is expensive, complicated and unlikely to happen for decades to come.

A simpler idea is to look for a second genesis right here on Earth. No planet is more Earth-like than Earth itself, so if life does arise readily in Earth-like conditions, then surely it should have formed many times over on our own planet. Well, how do we know it didn't? According to the orthodox picture, all life on Earth descended from a common origin, often expressed, following Darwin, by analogy with a tree. There is strong evidence that all life so far studied in detail is closely interrelated: those organisms use a universal genetic code, and they all employ nucleic acids to store information and proteins for structural and enzymatic functions. Proteins are always made by ribosomes. It is unlikely that so many specific features would have evolved independently from separate origins; rather, they were surely present in a common ancestral organism (often known as LUCA – discussed by Nick Lane in Chapter 11) and have been retained as frozen accidents. Even the so-called extremophiles – microbes that thrive in conditions that would be lethal to most life that we know – possess the same biochemical features, and share many genes with less exotic organisms. All known extremophiles have been positioned on the same tree of life as you and me.

Nevertheless, the vast majority of terrestrial species are

microbes, and biologists have only just scratched the surface of the microbial realm. Most microorganisms haven't been cultured or characterised, let alone genetically sequenced. At the present time, we simply don't know what they are. One cannot tell by looking whether a microbe is a bacterium or a novel organism with a radically different internal structure and biochemistry. To fully identify a microbe it is necessary to delve into its biochemical innards. It is therefore entirely possible that among the billions of microbes contained in, say, a sample of soil or seawater, some are representatives of life as we do not know it – 'weird life', as it is sometimes called, or 'Life 2'. And even if all microbes so far sampled are representatives of standard life, there may exist still unsampled niches that lie beyond the domain of comfort of even the hardiest extremophiles. Such niches could be inhabited by weird microorganisms.

Little attention has been paid to the possibility of weird (i.e. non-standard) life on Earth, although astrobiologists have thought a lot about weird life on other planets. Searches for weird terrestrial life fall into two categories. First is the case of ecological separation. Life 1 and Life 2 might inhabit non-overlapping regions or be restricted to different ranges of a particular parameter space such as temperature, pressure, etc. Consider, for example, hyper-thermophiles. The current upper temperature record is 122°C. If an exotic form of microbial life were detected in a deep-ocean volcanic vent system thriving at, say, 160–180°C, then it would stand out as a candidate for alternative life because of the discontinuity in the temperature range. Similar reasoning applies to life in strong ultraviolet regions such as the upper atmosphere and high plateaux, regions of extreme cold (Antarctica, mountain tops), aridity (Atacama desert), highly saline or high/low pH aqueous environments, heavily contaminated mining sites, and high-radiation environments such as uranium mines and nuclear waste deposits.

Much harder to identify would be weird microbes intermingled with standard life, especially at low relative density. Here, two approaches suggest themselves. One could devise a crude filter that would eliminate or at least inhibit the metabolism of standard life in the hope that it would leave any weird life unaffected. Then the weird life would eventually come to predominate. One hypothetical example of such a filter is a polymer customised to bind to some universal molecular feature of all life as we know it. If the polymer is loaded with a metallic nanoparticle and the system is then irradiated with a laser or microwaves, it would kill the host cells but leave any weird cells unscathed.

A second approach would be to make educated guesses about the nature of weird life. Synthetic biologists are trying to create novel forms of life in the laboratory, so they have become adept at imagining alternative ways that organisms could make a living. The problem about looking for life as we don't know it, however, is that we don't know quite what to look for. Any general signatures of life, such as carbon cycling or chiral specificity, will be masked by standard life. But if we guess that weird life might exploit a specific molecule, such as an amino acid not used in standard life, then methods could be devised to detect that molecule.

One clear-cut example is chiral specificity. Standard life uses left-handed amino acids and right-handed sugars. The laws of physics are, however, indifferent to the chiral signature of organic molecules (which way round they are), and a second genesis might well produce life with the opposite chirality, i.e., right-handed amino acids and/or left-handed sugars. A culture medium made of 'mirror molecules' would prove indigestible for standard life, but may be palatable to 'mirror' life.

If it can be established that life on Earth has originated more than once, it implies that life will emerge readily in Earth-like conditions, and it will therefore be very likely to arise on other Earth-like planets too. It is exceedingly unlikely that life would have happened twice on Earth but never on all other Earth-like planets. On the other hand, if weird life turns out to be merely a highly divergent outlier on the same tree as familiar life, we could not draw this momentous conclusion.

It is hard to imagine a discovery of greater significance to astrobiology than a shadow biosphere on Earth consisting of a different form of life descended from an independent origin. Obviously the existence of such a biosphere is a long shot, but it certainly cannot be ruled out on the basis of our current scientific understanding. If Earth has, or once had, a shadow biosphere, it would then be very likely that life will have arisen on many Earth-like planets across the Universe, as many astrobiologists currently assume, but with little or no justification. Because Earth-like planets are likely to be common, life could then be regarded as a truly cosmic phenomenon.

The philosophical ramifications of this shift in viewpoint are immense. So long as we know of only a single sample of life, it is possible to argue that biology is a freak local aberration, the product of a chemical fluke so improbable that it will not have happened anywhere else in the observable Universe. Although individual human beings may imbue their lives with significance, life as a whole would be a collection of insignificant freak physical systems, restricted to an infinitesimal patch of the cosmos. By contrast, if life is a 'cosmic imperative', emerging more or less automatically in myriad locations, we could say that the Universe possesses intrinsically biofriendly physical laws, so that life, and perhaps mind, could be regarded as having universal significance. They would be fundamental, as opposed to incidental, cosmic phenomena. One could then declare that we truly are at home in the Universe.

14

Alone in the Universe: The Improbability of Alien Civilisations

Matthew Cobb

When, in 1950, the physicist Enrico Fermi asked his famous question 'Where are they?' he was identifying the fundamental issue when it comes to thinking about aliens. There are billions of planets in our Galaxy alone, and even if only a very small proportion of them are Earth-like, that still leaves many worlds on which life could have taken hold. And yet we have no evidence of life elsewhere. The skies remain silent, space is not full of ingenious alien devices, and the few places our robots have visited appear barren.

The fundamental problem in trying to estimate the likelihood of life elsewhere in the Universe is that we know of only one kind, our own. As Francis Crick pointed out, life as we know it involves the flow of matter, the flow of energy, and the flow of information. We can imagine forms of life that follow this definition but are very different from our own – for example, non-cellular plasma-based life forms, or huge single-cell organisms, or even life living in two dimensions, or indeed aliens living in parallel Universes. However, we have no evidence for any of this, beyond the power of our imagination and the suspicion voiced by the evolutionary geneticist J. B. S. Haldane that 'the Universe is not only queerer than we suppose, but queerer than we can suppose'.

Alien life, queer or not, would have to conform to the laws of physics. This could lead us to encounter some familiar forms – if there are metre-scale fast-moving predators in the seas of Enceladus (or in any other liquid anywhere in the Universe), they will have a profile something like that of a shark or a squid. These organisms have similar shapes because of convergent evolution: the physics of moving through liquid requires that fast organisms be streamlined. However, this does not mean that everything about life on Earth will be repeated elsewhere.

In the first version of his equation, Frank Drake suggested that all planets that could support life would do so, and that all planets that supported life would go on to develop intelligent life. The reality of life on Earth – the only example we have – indicates that such high probabilities are profoundly mistaken, and reveal a fundamental misunderstanding of how evolution takes place.

We can be beguiled by our unique abilities, and indeed by the very fact of our existence, into imagining that our evolution was the expression of evolutionary trends towards increased intelligence, and that given the immensity of space, these tendencies will be repeated on other worlds. None of this is true. There is no direction to evolution, and there was nothing inevitable about our emergence as a conscious spacefaring species. All major evolutionary changes occur in response to environmental challenges that are contingent and unpredictable. Those changes alter the ecosystem and lay the basis for new developments – for example, the evolution of multicellularity – but without the largely random events that underpinned those changes, life on Earth would have been very different indeed.

A study of the key points in life's history on Earth – the only kind we have any knowledge of – leads us to the conclusion that the answer to Fermi's paradox is that its starting point is probably wrong. There are no alien civilisations.

Abiogenesis

The events leading to abiogenesis – the appearance of living matter from non-living components – are unknown. There is no scientific consensus over the most likely scenario – experimental evidence may eventually support one of the competing hypotheses, but that seems some way off. It appears that life began almost as soon as conditions on Earth were appropriate. A few hundred million years after forming, Earth became host to organisms that were the ancestors of all forms of life we see today. This could imply that abiogenesis can occur relatively easily, although this is not a logical conclusion. Because we do not know the conditions that led to abiogenesis, we cannot calculate the probability of them occurring elsewhere. If those conditions turn out to have been incredibly specific and unlikely, then even with vast numbers of potential planetary hosts, it may be that our form of life is the only one that exists.

If abiogenesis is straightforward, we need to explain why we have no evidence that it has happened more than once on Earth – we know that all existing life has a common ancestor because of similarities in our DNA, while no other form of life has been detected. Darwin's solution to this enigma was that any newly appearing forms of life would simply be eaten before they could take hold. This may be the best answer there is, but there have been 3.8 billion years for life to appear on Earth a second time, and yet it apparently has not done so. Abiogenesis may have only occurred once in 3.8 billion years because it is highly improbable.

Even if we accept that abiogenesis is a relatively trivial event, that would almost certainly mean that we live in a Universe of slime, populated at best by unicellular biofilms aggregating on the surfaces of exoplanets. Once life appeared, the four preconditions for the existence of a spacefaring species on Earth – the eukaryotic cell (defined below), multicellularity, self-awareness and civilisation – were each extremely unlikely and apparently unpredictable.

They were certainly not inevitable, given the existence of life. Multiplying these improbabilities strongly suggests that we may indeed be alone!

Eukaryogenesis

All complex multicellular organisms on Earth are what are known as eukaryotes: they have complex cellular structures including a nucleus that houses the chromosomes, organelles for synthesising proteins and above all mitochondria – small energy-producing structures that enable eukaryotic cells to become up to a million times larger than cells without mitochondria, and can also allow for cells to combine to form multicellular organisms. We can imagine alien life forms without mitochondria – there are trillions of microbes like this on Earth – but for alien life to outgrow the microscopic, it would have to generate vast amounts of energy to power large-scale organic structures. Strikingly, we have no evidence from Earth that evolution by natural selection is able to find such a solution – in the nearly 4 billion years life has had to come up with an answer, natural selection failed.

What happened on Earth – known as eukaryogenesis – was not the consequence of random mutation and the subsequent sifting of inherited characters that have differential fitness (the essence of natural selection). Instead there appears to have been a single event of mind-boggling improbability, for it involved two life forms interacting in a most unusual way.

DNA evidence shows that this took place only once in the history of our planet, on a Tuesday afternoon around 2 billion years ago, somewhere in the ocean. Prior to that moment, all life had consisted of small microbes with no cell nucleus or mitochondria. Everything changed when one unicellular life form, known as an archaebacterium, ended up inside another, called a eubacterium. It seems probable that by combining their metabolisms, these two very different organisms gained an advantage by being able to exploit novel food resources. But although this may have initially been an equal relationship, the eubacterium was trapped, and over millions of years and billions of rounds of unicellular replication it lost many of its genes to its host and eventually became a slave, a mere molecular powerhouse – the mitochondrion – that produced energy from chemical reactions and was used by the new eukaryotic cell. Armed with this novel source of energy, these new eukaryotic life forms went on to prosper, as natural selection slowly began to play with the new creation.

We could in principle calculate the probability of the appearance of eukaryotes, but we would soon run out of zeros. Here's why. Right now, there are more single-celled organisms alive on Earth than there are Earth-like planets in the observable Universe; the total number of single-celled organisms that have lived on our planet over the last 3.8 billion years is incalculable, while the number of times they all interacted is even higher. And yet out of all those gazillions of interactions, just one of them created a weird hybrid in which one of the partners eventually became first a trapped symbiont, and ultimately merely an organelle, providing the larger organism with energy.

That weird hybrid was our ancestor, and its existence – and therefore ours – was incredibly improbable. As far as we are aware, no such event happened before or since. Given that we know this was an incredibly rare and entirely chance event, there can be no certainty that anything like this has happened on any other planet in the history of the Universe.

Perhaps this view is too pessimistic: after all, around a billion years ago, something vaguely similar happened again and another form of symbiosis eventually appeared when a eubacterium that had evolved the trick of acquiring energy from sunlight found itself inside a eukaryote, complete with its mitochondria. Like eukaryogenesis, this event happened only once in all the vast history of life and gave rise to algae and eventually plants, in which small organelles called chloroplasts, the descendants of the eubacterium, turn light into energy for the benefit of the eukaryotic host. While a second occurrence of such weird symbiosis-turned-hybridisation at least doubles the likelihood of such an event, it still leaves it in the realms of the highly improbable. Alien life would need to have a similar range of organisms, with similar habits, before such an event could even occur. That makes it incredibly improbable that the kinds of organisms we see on Earth have their equivalents elsewhere.

Any large alien life form would require some way of transporting matter, energy and information from the environment into its interior. Life on Earth without mitochondria is limited to the microscopic because of the physical limits imposed on that transport in the absence of an additional powerful energy source. The co-option of the energy-producing mitochondria first enabled eukaryotic cells to grow large, and then, eventually, to become multicellular. But without the luck of acquiring a symbiont that could provide the necessary energy and could eventually be entirely subsumed, life could not escape the minute. If there are aliens, it is most likely that hundreds of thousands of them could fit on the head of a pin.

Multicellularity

All the multicellular life forms we are so familiar with can trace their ancestry back to that single event that created the eukaryotes, but this does not mean that the appearance of multicellular life was inevitable. For well over a billion years after eukaryogenesis, life continued to be resolutely unicellular – indeed, most eukaryotic lineages are still unicellular today. Because conditions on the planet remained basically unaltered, life could gain no advantage by changing. As a result, Earth looked pretty much the same for nearly 3 billion years – there was no life on land, and apart from the odd oceanic algal bloom and the presence of lumps of stone produced by sand particles trapped in bacterial mats, there would have been nothing to indicate to any passing aliens that anything interesting was going on down here.

The lesson is that there is no evolutionary drive to multicellularity. Indeed, there are no evolutionary drives at all, except, perhaps, to live and to reproduce. Multicellularity eventually evolved perhaps 25 times in four major lineages – animals, plants, fungi and brown algae – but as with abiogenesis we do not know how, why, or even precisely when this occurred. Genetic data tracking the similarities between animals suggest some lineages may have separated 700 million years ago, while other multicellullar life forms evolved earlier.

The reason life took the multicellular turn must have been due to changes in the environment, linked with climate change or geological shifts, combined with mutations that occurred repeatedly within the original eukaryotic lineage, perhaps leading to cumulative changes that eventually flowered when the world altered. In this new world, larger and more complex life forms could survive and prosper; as they did so they began to shape the environment directly, burrowing into the bacterial mats on the sea floor, churning up the substrate, creating new ecosystems. With changes in the availability of oceanic biominerals at the beginning of the Cambrian period (542 million years ago) – perhaps following the action of glaciers that covered the Earth and linked with increased oxygen levels – animals could develop shells and then rigid exoskeletons for movement and defence. A kind of evolutionary ratchet or arms race began to function, giving the appearance of progress and leading to the evolution of heightened senses all round, increasingly rapid and cunning predators and their counterpart – skittish prey that would shoal and produce collective behaviour to avoid predation. The exponential wonder of natural selection began to operate, creating an amazing variety of animals during the aptly named Cambrian explosion. This was not the teleological expression of some inner drive of life, but the consequence of environmental and genetic changes and the effects of the many arms races that appeared. However, to get to this point required an incredible number of unlikely events, both biological and geological. What followed was equally improbable.

Near misses

Even once the land had started to turn green and the oceans were full of amazing life forms, the path from then to now was neither inevitable nor straightforward. Chance events shaped the way our planet evolved, and sometimes there were some very narrow escapes. For example, at the end of the Permian period, around 252 million years ago, 90 per cent of marine species and around 70 per cent of terrestrial species went extinct as a wave of gigantic volcanic eruptions changed the climate for millennia. Our ancestors survived, but it is not difficult to imagine that they might not have done had things been slightly different.

The most well-known mass extinction event, which got rid of the non-avian dinosaurs, was at least partly caused by a massive meteorite impact 66 million years ago. Had some aspect of celestial dynamics been only slightly different, the asteroid would have missed the Earth, and you and I would not be here, for there would have been no dinosaur-free ecological niche into which our mammalian ancestors could have expanded. Chance events like this have shaped our planet, and will have shaped life on other planets, too, if it exists. We just need to remember that 'shape' can mean 'destroy'.

On an alternative Earth that continued to be dominated by dinosaurs, we cannot assume that some kind of highly intelligent reptile would have evolved in our place. There is no evolutionary tendency for animals to become more intelligent, or more complex. For example, tool use is seen in a variety of animals, including some birds. It is possible that this habit goes way back in evolutionary time, when birds were simply another kind of dinosaur. This impressive ability has not led to the planet being dominated by crows. Only one species has turned tool use into global domination: ourselves.

Self-awareness and civilisation

Just as there was no evolutionary drive to complexity and multicellularity, there was no evolutionary drive to consciousness. Instead, a series of evolutionary changes in response to random events took a tortuous, complex and highly contingent route, ending up with me transmitting my thoughts to you through the medium of the written word. There was nothing inevitable about that path.

There is no scientific way of defining which animals are conscious – our impressions about animals' facial expressions and behaviour are not a good guide! If we assume that the great apes – chimps, orangutans and gorillas – are somehow conscious in a way similar to us, and that our common ancestor was too, that suggests that consciousness has been around on our planet for about 10 million years. If we include other lineages, such as some large mammals (whales or elephants), or perhaps crows, then that could suggest that it has been around for longer, and presumably evolved more than once. However, there is a qualitative difference between human consciousness and that shown by our ape relatives, never mind the more contested forms that might exist in other mammals, or even in some birds. We can speak, read others' thoughts, imagine what others think, and tell lies. No other animal can do all these things. As far as we know, our consciousness, our way of thinking, is unique in the history of Earth.

The appearance of humans in East Africa, with our limpid self-awareness and our ability to think in complex abstract ways, and express those ideas through language, was not inevitable. It was the result of another series of chances, probably related to climate change. Anatomically and psychologically modern humans first appeared at most 200,000 years ago – this means that for about 99.995 per cent of the period for which life has been around, there has been nothing an alien could talk to. The occupants of a flying saucer might have been impressed by Earth's stegosaurs, sharks and slugs, but if they wanted to pass on the secrets of their superior technology, they would soon have flown off elsewhere.

Even once we appeared, our survival was by no means certain: genetic data show that the human population went through a near-catastrophic crash about 80,000 years ago – at one point there were only around 10,000 people on the planet. The fragile flame of humanity could easily have been snuffed out by drought or disease. Once we began slowly spreading across the globe, chatting as we went, doing ceremonial dances, tattooing our bodies, and painting on cave walls, we met close relatives such as the Neanderthals, the mysterious Denisovans and probably others. We have survived, they have not; but it could easily have been otherwise.

Even after we had spread across the planet, expanding our population to a million or so, there was still no guarantee that the talking ape would make it into space. Civilisation required the conjunction of plants that could be domesticated and an appropriate climate; we were able to exploit these conditions, first in the Fertile Crescent (modern-day Iraq and Syria), and then in China and Central America, but the earliest experiments in agriculture could easily have failed. For most of our existence we have been hunter-gatherers and without the development of agriculture this would have continued. Yet again, the course of our development was paved with luck, and the chances of success were very slim. Conscious aliens would be subject to the same rigours of chance, and they might not be as lucky as we have been.

With luck and a great deal of judgement we will overcome the existential threats of climate change and nuclear annihilation and we will avoid future zoonotic epidemics and the menace of antibiotics rendered ineffective by overprescribing, and our space alert systems will enable us to avoid a future massive incoming asteroid (these are just the dangers we can imagine; there may be others). If we manage this, we may colonise other worlds and we could send out thousands of probes, inviting contact with alien civilisations. But in the end we will go extinct, as all species do. The temporal window within which other civilisations could identify and contact us is likely to be extremely narrow; on a galactic timescale we will not be around for long. The same could apply to any hypothetical aliens. Tragically, we might simply miss each other. Although there is the possibility that we will encounter their probes, relics of a dead civilisation, for the moment we see no robots apart from our own.

The fact that we have made it this far does not imply that there must also be spacefaring aliens, nor that we are somehow destined to reach the stars. The apparent inevitability of the existence of human civilisation is a trick of perspective, a cosmic tautology: we can only wonder about such matters because we are here. Our existence has not been guided by some supernatural force, nor is it written in our genes. We have just been very, very lucky.

This view of life is not bleak; it is simply the way things are. It does not suggest that SETI or similar projects are a waste of time, nor that we should not explore our solar system and beyond. But it does imply that our priority should be to understand the fragile wonders of life on Earth and do our best to ensure that we do not damage the ecosystem even further. We have a responsibility to the trillions of organisms that live on Earth, whose existence we have severely disrupted and threatened. Solving the immense problems we have unwittingly created should be our main target.

No one would be happier than me were we to find non–Earth-based life on Mars, or if we were to pick up a message from the stars. I would be overjoyed to be wrong, but I am not holding my breath.

ALIEN HUNTING

15

It Came from Beyond the Silver Screen! Aliens in the Movies

Adam Rutherford

They mostly get it wrong.

Mostly.

Film-makers have been infusing culture with their visions of aliens for more than a century, and almost all of them have been a lot like us. The Moon natives in the first cinematic trip into space, Georges Méliès's *La Voyage dans la Lune* (1902), were Selenites, named after Selene, the Greek goddess of the Moon. They're a bit like arthropods with bulbous heads and lobster claws, but mostly human – upright and bipedal. The next trip was when the 1919 adaptation of H. G. Wells's *The First Men in the Moon* landed, which also had Selenites as the endogenous lunar men. Alas, all prints of the film are lost. In the few remaining stills from the shoot, the Selenites are also somewhat insectoid, but look disturbingly like the blue, globoid-headed, oval-bodied Igglepiggle from the bewildering otherworldly toddlers' programme *In the Night Garden*.

And so the tone was set for a century of aliens – humanoid, insect or insect-like humans are the mainstay of cinematic extraterrestrials. We turn to human-like forms either because of budgetary constraints or for reasons of anthropocentrism.

We lazily assume aliens will be a bit like us, because we do like thinking about ourselves. *Star Trek* and dozens of imitators have got away with simply gluing bits of lump onto human faces or painting them green to indicate their non-human status. The *Star Wars* Universe offers little but variations on humans. Budget didn't seem to be much of a problem in James Cameron's *Avatar* (2009), just a tiresome lack of imagination. 'Let's make them taller than us, and a bit cat-like, but sexy, and give them tails. They need to be primitive but wise. Oh, and make them blue too.'

We have a pretty good grasp of evolution these days, and our bounteous fossil record, now coupled with genetics, gives us a picture of how life evolved on Earth. There are plenty of mysteries remaining, but we know much about our nearest ancestors: the emergence of bipedalism and all the many factors by which we came to be who we are. To assume that on other worlds, evolution would deliver a species identical in physical stature is plain silly. We don't really know why we became two-legged when almost all terrestrial animals are not, but we can hypothesise that it is an adaptation to a range of complex environmental conditions, primarily to equip a species for a life on the savanna rather than swinging in the trees, and an increased efficiency of movement. If the Earth ever got a reboot, and the story ran again from the beginning, with just a few variables altered we would not have come out like this. Even a seemingly unconnected matter like the tilt of the Earth's axis has played a crucial role. That 23° tilt, which gives us our seasons, was caused by a rock the size of Mars colliding with the neonate Earth, and knocking off a block that would form the Moon. Imagine if the rock had missed; no tilt, no seasons, no Moon, no tides. This would have meant a different weather system, different climate changes over time, and an entirely different set of evolutionary ancestors. Imagine if that six-mile-wide asteroid hadn't tumbled out of the Cretaceous sky into what is now the Gulf of Mexico and caused an extinction level event that wiped out the dinosaurs and so many other species, yet allowed our small mammal ancestors to thrive. Imagine that rock being half the size, and only half of the dinosaurs had been wiped out. Would we be as we are? The answer is almost certainly no. Our form is not inevitable – it's mere cosmic happenstance.

Since the 1950s, aliens have frequently appeared as indistinguishable from humans on screen. Sometimes this is simply filmmaking on the cheap. Ed Wood's *Plan 9 From Outer Space* (1959) is an equally derided and loved cult movie, sometimes cited as the worst ever made, with comically wobbly sets, flying saucers on strings, a clumsy and prolix script, and a cast of wrestlers, local celebrities and a female vampire impressionist. Human aliens arrive with their infamous 'Plan 9' to resurrect the Earth's dead, notably in the form of Bela Lugosi (who, Fact Fans, died a few days into filming, and was replaced by Ed Wood's chiropractor, who was much taller and bore no resemblance to Lugosi, but instead held a cape in front of his face; it wasn't even the same cape though, as Lugosi was buried in the original). Eight years earlier we met Klaatu, the humanoid in the classic *The Day the Earth Stood Still* (1951), with his atomic-age warning to Earth that we better shape up or be 'reduced to a burned-out cinder'. There are also plenty of representations of an assumed evolved human alien, the so-called Greys, with spindly body, and big brainy heads and eyes, supposedly implying their superiority to us through cerebral evolution, and the move away from physical brute. Steven Spielberg seems to like them particularly. We see them in *Close Encounters of the Third Kind* (1977), back lit, gracile and graceful, and in fact played by young girls. *E.T.* (1982) was more green than grey (until he died), but those bodies return, even thinner and bulb-headed in *AI: Artificial Intelligence* (2001), though these turned out not to be aliens, but super-evolved robots.

We humans represent a very small proportion of life on Earth. Most organisms are single celled – bacteria or archaea, but I guess they're too small to make cinematic creatures (though off-screen, even smaller replicating entities, viruses, do play the decisive role in our survival against the three-legged Martians from the *War of the Worlds*, 2005). Most animals on Earth are insects. We share ancestors with those creatures that crawleth about 550 million years ago, and though insects look very different from us mammals, the genes responsible for the formation of legs and eyes and the whole axis of their bodies are pretty much the same as in mammals, and their bodies are structured in the same way – a head with eyes and a mouth at one end, a tail at the other, legs in the middle. Nevertheless, since the Selenites, insects and other arthropods have served as inspiration for otherworldliness: *Starship Troopers* (1997), *Independence Day* (1996), *Men in Black* (1997), *District 9* (2007), *The Mist* (2007), and dozens more.

In 1979, we were introduced to a new extraterrestrial insectoid in *Alien*, and then a whole swarm of them and their big-assed mother in 1986 in its sequel, *Aliens*. A 7 ft 2 Nigerian actor called Bolaji Badejo was inside that xenomorph suit in the first film, and there is a razor-tipped tail, and a lot of phallic imagery, but it is still a man in a suit – head and mouth(s) at the top; arms, fingers and legs all the way down. By *Alien 3*, the parasite has infected a dog, and the resulting mature animal is canine rather than humanoid. For all of the anthropomorphism of *Alien* and *Aliens*, and the caninomorphism of *Alien 3*, these creatures fit into a scientific idea that has been well considered. These titular xenomorphs are parasites, and their behaviour is perfectly believable in light of some of the breathtakingly grim parasitic behaviour we see in nature. If you don't believe me, here are some examples.

The Alcon Blue butterfly (*Phengaris alcon*) is very pretty, but appearances can be deceiving for it is a rather wicked creature. It lays its eggs on the Swiss wild flower *Gentiana*, where the larva feed until they are fatted, and then loll on the ground waiting to be discovered by ants. The grubs secrete a chemical which mesmerises the poor deluded ants into thinking they are their own babies, and bring them into the hive, whereupon the butterfly grubs eat the ant grubs. Once ready to emerge into the world from its ant-cuckoo brood, the butterfly does have to run the gauntlet to escape, as the ants suddenly realise that this flap-winged thing is not actually one of them at all. However, the newborn butterfly is armoured with flaky scales that the ants struggle to grab hold of, and it bludgeons its way out, hotly pursued by some irate cuckolds.

And if you think all this evolved opportunism is remarkable, consider the wasp *Ichneumon eumerus*: its main hosts are Alcon Blue grubs! The females scour the ground for the scent of the ant colonies, and will only enter those that have the butterfly larvae in them. Inside, she pierces the belly of the fattest butterfly grub using her very pointy ovipositor, and inserts a single egg. She also marks the nest with a chemical that warns off other ichneumon wasps from doing the same. After nine or ten months of being nurtured by ants inside a butterfly maggot pretending to be an ant, the wasp is ready to burst from its host, and releases a chemical that causes the ants to fight each other and not attack the wasp.

Parasitism like this is very alien to us humans, and yet it abounds in nature, and it's pleasing to see elements of a parasitic life cycle present in the *Alien* films; the insertion of an egg into a host; the messy bursting forth; the armour plating; the shed skin. But imagine pitching the story of the Alcon Blue to a Hollywood producer. Nature is frequently hard to believe, and this butterfly story sounds just a little unlikely. Alongside the original *Alien* films (of which there were eventually four), there were two horrid spin-offs featuring another filmic alien, the Predator. The best thing about these wretched films was the tagline for the *Alien versus Predator* poster: 'Whoever wins ... we lose'. That's how the ants must feel.

It is possible to get too lost and obsessed with the scientific verisimilitude in films. In general, I'm not that bothered if the science is not watertight, or set in galaxies far far from reality. One alien film, though, has caused me anger, hate and suffering – and, as all science fiction fans know, these emotions only lead to the dark side. *Prometheus* was so inconceivably awful in every aspect, and made so little sense in terms of basic plotting, that it seems fair game to pile criticism onto the woefully ill-conceived science, which was presumably intended to underwrite the plot, but in fact undermined it. It begins with an extremely tall and preposterously muscled human-like alabaster-skinned man wearing a nappy on a clifftop, possibly in Iceland. There, for reasons unclear at the time, and never explained, he consumes a small vial of black stuff, grimaces, and then crumbles fragmented into the water below. The camera zooms into the molecules of his disintegrated life.

As the title card shamelessly mocked the iconic lettering of the first *Alien* film, DNA from the giant muscular fellow swirled in the waters of this primordial incubator. But it was a double helix that twisted to the left, and thus *Prometheus* was wrong from the very first frame. All DNA on Earth turns to the right, like a corkscrew, a fact that betrays its singular origin and the shared ancestry of all life on this planet. This mistake happens a lot, and is sort of forgivable if I'm in a good mood. But it's just wrong. Anyway, with this set up, the idea is that we were engineered by a species that came before all life on Earth and gave up its DNA (regardless of its handedness) to seed evolution. It's a version of a theory called panspermia, much loved by science fiction authors, where aliens plant life on other planets, either through intent or just accidental cosmic pollution. It's a nice idea, but in my view firmly in the realm of science fiction as we have good models for how life began on Earth, and none of them requires either alien or divine intervention. According to the plot in *Prometheus*, these ancient anthropomorphised aliens gave the planet DNA, which over the course of life on Earth ended up with us looking like shorter, darker and altogether less beach-bodied versions of our ancestors. If that were the case, then why did evolution take such a circuitous route to get back to the creator's form? Why did we spend so much time hairy and on all fours? Or as brutish reptiles? Or indeed, why did our ancestors spend so much time – probably 2 billion years – as single-celled organisms, if we were only waiting to become lesser versions of what our creators already were? This panspermia fallacy is not the only alien encounter in *Prometheus*; there's a much more phallic one too. One of the mission scientists discovers a wormy alien species. This, we assume, is the first contact that any human has made with an extraterrestrial, and because he's probably the worst scientist in the history of science, he immediately takes his helmet off. He then does a sort of 'coochy coochy coo', as the worm rears up to reveal not just phallic imagery but a fanged vagina as a mouth. It then chews his face off, which comes as some relief to the audience (though it should be noted that such is the baffling narrative and scientific confusion in this film, this doesn't kill him; he comes back later as a giant, angry, blister-headed zombie – perhaps this is how anyone would react to having one's face devoured by an acid-spitting *vagina dentata*).

I digress. *Prometheus* is a terrible film because of its nonsensical script, not because of its lazy science and unimaginative aliens. It is impossible to get aliens right on screen, because we have a sample size of life in the Universe of just one. For all the astonishing variety of life on Earth, it is all part of the same tree. We have the same DNA (which is always a right-handed twist), the same cell structures, the same basis for harvesting energy from the environment.

I consider that there are two ways of creating successful aliens on screen. The first is not to try at all. Disguise is a perpetual theme of humanoid aliens. They walk among us, hiding either to enact their nefarious plans (*Invasion of the Body Snatchers*, 1956, 1978; *They Live*, 1988; *Under the Skin*, 2013), to fit in (Superman in many movies; *Starman*, 1984), or simply to survive (*The Thing*, 1981; *The Man Who Fell to Earth*, 1976, though it has never been made explicitly clear that the actor who played the title role in the latter film, David Bowie, was terrestrial in origin. Wherever he is now, it's not Earth.)

The other method of dealing with an authentic alien is simply to be incomprehensible. Stanislaw Lem wrote the science fiction novel *Solaris* with this in mind.

> [I] wanted to create a vision of a human encounter with something that certainly exists, in a mighty manner perhaps, but cannot be reduced to human concepts, ideas or images.

It was filmed three times, in 1968, 1972 and 2002. The two most recent are both stunning in different ways, both considered meditations on death and consciousness. The non-human life is the planet Solaris, as far as we can understand it, around which a space station orbits, with a crew in place to study it. The alien presence manifests itself as memories of people. Dead relatives or wives are conjured from the minds of the crew, incomplete or misremembered, and always upsetting, yet addictive to the extent that the crew don't want to return to Earth, even as their ship's orbit decays into the planet. No attempt is made in either film to explain the alien intelligence; it is merely an expression of consciousness utterly different from our own.

Kubrick's science fiction masterpiece, *2001: A Space Odyssey* (1968), uses a similar idea, with the presence of the black monoliths, one triggering the advent of weaponised tool use in our ape ancestors, and another 3 million years later propelling us forward in an unexplained evolutionary leap. In doing so, he upturns Darwinian natural selection in a way that does not upset me, an evolutionary biologist. What is the alien? We don't know. But it is not us, nor anything we recognise.

Carl Sagan's novel *Contact* was made into a movie of the same name in 1997 starring Jodie Foster as the astronomer Dr Ellie Arroway, possibly the best on-screen scientist in cinema's history. She detects a repeating signal from a nearby star system that cannot be anything other than the product of extraterrestrial intelligence. Within the signal there is a set of instructions on how to meet the entities at the other end.

This is perfect science fiction, in that it is fiction rooted in science. It is a fantasy set-up and it relies on non-existent technology and phenomena (such as the Hollywood staple for travelling inter-stellar distances, the wormhole, which are at best theoretical). But they are in the service of a fantastic and skilfully told thriller about how science works and how and why we explore.

The alien tech works. When Dr Arroway awakes to find herself in a distant star system, she's on a tropical beach. The sky is different, the wormhole swirls above her, and it all shimmers in an unearthly way. A blurry spectre approaches, and as it comes into focus, it is not alien at all. It is her father, who had died when she was 10. Some people groan at this, writing it off as base sentimen-tality. I find it powerfully moving, not least because she figures it out in seconds. 'None of this is real ... While I was unconscious you downloaded my thoughts, my memories even?'

'There's my scientist,' the alien replies. 'We thought this might make it easier for you'. Carl Sagan figured it out too. We can't conceive of alien life. If it does exist, it's difficult to imagine it not being Darwinian in its nature, and similarly it's hard to imagine that it wouldn't work in a similar way to life on Earth, that is, powered by a mechanism for continuously extracting energy from its surroundings, and slowing down the inevitable increase in entropy as long as it lives. But we cannot imagine the evolutionary pressures that an alien would have endured over billions of years to shape its physical form, or its behaviour. If there is intelligent life in the Universe, we will have to wait a long time before we meet it. In the mean time, we search. Are we alone? Other contributors in this collection will have given reasoned or mathematical answers. For me, the real answer is that the more we look, the more we find out about ourselves, both in science and in science fiction. As *Contact*'s alien concludes (though I can quite imagine Carl Sagan saying it himself):

> In all our searching, the only thing we've found that makes the emptiness bearable is each other.

16

What Are We Looking For? An Overview of the Search for Extraterrestrials

Nathalie A. Cabrol

Over fifty years ago, SETI Institute astronomer Frank Drake wrote the now famous *Drake equation*, becoming the first scientist to articulate a global vision for the search for life beyond Earth. In doing so, Drake was establishing a 'roadmap', a route forward which demands that, if we are to successfully search for life in the Universe, we must first understand how galaxies and planetary systems are formed, how many planets may become habitable and may develop life, civilisation and technology, and how many of those advanced civilisations might want to communicate with us. Although his formula was directed to searching for advanced extraterrestrial civilisations, it already included most elements of modern astrobiology. The Drake equation shows how approaching the question of the origin of life and its potential existence beyond Earth requires a holistic approach, the same vision embraced today by astrobiology. At the crossroads of scientific disciplines, astrobiology uses advances in all fields to answer these questions: How does life begin and develop? Does life exist elsewhere in the Universe? What is life's future on Earth and beyond?

These questions represent a puzzle of cosmic proportions, to which we are missing several key pieces. We do not have a clear definition of what life is. Could it have been seeded on Earth through panspermia (in which comets and asteroids transfer material between other bodies in the solar system on impact) and planetary exchange (the idea, for example, that there was some exchange of material between Mars and Earth at the time they were forming)? Or was it created on our planet through abiogenesis, a process by which life arises naturally from simple organic compounds and chemical processes? We also do not have a record of when – or in which environment – the transition from prebiotic chemistry to life took place. We don't know whether life is a common universal occurrence or a fluke. But if we are to solve the puzzle, it makes sense to start with *us*.

The terrestrial biosphere we inhabit – even if it hasn't provided the answers to the questions above – is a record of life's evolution and adaptation driven by environmental and cosmic bottlenecks, extending over billions of years. Further away, we can see the solar system we belong to as a lab where, over eons, nature has created a diversity of environments surpassing in complexity anything we could develop in an experiment. Beyond the solar system, our most sophisticated instruments provide windows in space and time where we can catch a glimpse of how galaxies, stars and planets are formed. Last, but not least, the human mind can model, theorise, and generate limitless thought experiments.

With this in hand, we have started to build an understanding of what, where and how to search for life beyond our planet. By necessity, our vision is still anthropocentric: we are searching for life *as we know it*, and this approach is a logical one because it is always easier to start with what you know, when what we know of life is still so limited. As our knowledge broadens, hypotheses and models grow more complex, and the technology to test them becomes more sophisticated, which allows more discoveries to be made, and fundamental hypotheses and models to be refined. This is an iterative process. In that regard, the past few decades of exploration of the Earth's most extreme environments, the solar system, and deep space have revolutionised our definition of habitability and life potential.

Life: what are we searching for?

As you will have discovered in previous chapters, the first challenge to finding life in the Universe is the absence of a broadly accepted definition of what life *is*. Biologists, biochemists and geneticists still have difficulties today agreeing on a unified definition, while some argue that we should not even try, because no single property definitely characterises the living from the non-living. However, one particular attempt to define life does provide a useful avenue for exploration: in 1944, Erwin Schrödinger proposed that living matter is that which 'avoids the decay into equilibrium' – or at least delays it by fighting entropy, the spreading out of energy towards a state of uniformity. As long as metabolic activity continues through biochemical processes like nutrition and the elimination of waste by-products, biological activity is being maintained – or, to put it another way, life is being lived.

Leaving aside the question of whether Schrödinger's definition is a direct observation of what life *does*, but not necessarily what life *is*, if metabolic activity is a way of observing and measuring life, then this is something that can be used to search for biosignatures beyond Earth. This is a technique which has already been used on Mars by the Viking missions (admittedly with debated conclusions).

In the case of exoplanets, the distances limit us to remote-sensing detection techniques, but we are starting to learn how to recognise the spectral signatures of life by studying the light from chemical compounds in exoplanets' atmospheres that could have been produced by living organisms. (For more on this, see Chapter 17.) The difficulty with this approach is that many of these gases, such as methane and oxygen, are not unambiguous signs of life, and can be the results of geological and/or biological processes.

Biomineralisation, in which living creatures produce minerals, often to stiffen their own tissues, will not be accessible to remote study via telescopes any time soon and therefore requires 'on-site' techniques. Once again, progress in the remote identification of biosignatures will come with a better understanding of what life is, and what its processes and by-products are, here on Earth – at least to start with, and we should not neglect the importance of our planet and solar system as training grounds as we begin our spacefaring explorations.

Life: where did it come from?

At the moment, our best guess is that life originated from simple organic compounds based on six important elements: carbon, hydrogen, nitrogen, oxygen, phosphorus and sulphur. This means that there had to be a transition point when chemistry became biology. Whether it is this transformative moment that *defines* life, or whether life is the *result* of this transformation and the many forms it subsequently took, is a question that remains unanswered, and science, philosophy and religion provide ways to approach it. Clues about how and where this transition from chemistry to biology happened could theoretically be located in the geological record of our planet. Until recently, the first indirect evidence of life was the 3.7 billion-year-old graphite discovered in Western Greenland. Microscale carbon isotope and laser Raman spectroscopy on the graphite indicated that the source of the carbon was biogenic organic matter: that is, created by living organisms. A recent study proposes that biogenic carbon could be present in 4.1 billion-year-old rocks in Western Australia. The first evidence of life in the Archaean era (4–2.5 billion years ago) in the form of stromatolites and microfossils was discovered in 3.48 billion-year-old sandstone in Western Australia. Stromatolites are formed when microorganisms (blue-green algae) in shallow water trap and bind together sedimentary grains to create layered rocky structures.

This evidence was found in what remains of the oldest terrestrial rocks. But such rocks are rare. They are barely a few hundred million years older than the original crust of our planet, and formed right after the Earth had cooled down. Unfortunately, most of the geological record from these early times has been recycled through erosion and plate tectonics. It may well be, then, that the record of the transition from chemistry to biology is gone forever and we might never be able to find it on our planet. However, it might still be present elsewhere in the solar system by virtue of planetary exchange between the early Earth and Mars.

Life: where to look?

In order for life to emerge and thrive, certain factors must be present: water, energy, nutrients, and protection from elements like intense solar or cosmic radiation which would threaten it. It has been argued that life appeared the moment our planet's environment was stable enough to support it. As we can see from the above, the first indirect evidence for life activity could be as old as 4.1 billion years, and interestingly, during that period Earth was still being bombarded regularly by massive asteroids, meteors and comets that would have churned up the surface down to substantial depths and profoundly disturbed the climate globally. This is also the same time frame suggested for the formation of the first oceans (4.2–3.8 billion years ago depending on models). Because of their depth, oceans could have provided a protective and stable environment over long duration for biochemistry to develop, possibly around hydrothermal vents.

These hydrothermal vents would have been extreme environments, with any life evolving there capable of withstanding extreme pressure: what we would now call 'extremophiles', creatures capable of living in extreme environments. If the earliest forms of life survived there – in fact, forming the origin of all life on Earth – then it bodes well for our search, because water, energy, nutrients and shelters are present (simultaneously) on a number of planets and moons of the solar system, albeit ones that have distinctly less comfortable environments than Earth. Our generation is just starting to grasp the amazing diversity of potential habitats for life in the solar system, but in barely fifty years we have made incredible strides in transforming the collective image of cold desolate planets and moons orbiting the Sun into that of exciting worlds full of possibilities where life could be the rule and not the exception.

At the same time, space and ground telescopes are already revealing glimpses of vastly different worlds far away from us. There are multiple exoplanets – planets outside our own solar system – which might fulfil the conditions needed for life. In 1992, the first hot Jupiter was discovered orbiting pulsar PSR 1257, about 2300 light years from our Sun. Three years later, 51 Pegasi b became the first planet found in the system around a main sequence star. Since that time, thousands of exoplanet candidates have been identified. Many are too close to their parent stars and perpetually in a molten phase. A large number are gas giants bigger than Jupiter while others are frozen solid. But among the thousands of candidates identified and confirmed, a few dozen Earth-sized planets and larger 'super-Earths' located in the habitable zone of their stars have drawn attention. Currently, three front-runners are: Kepler-186f, an Earth-sized exoplanet orbiting a red dwarf star nearly 500 light years away; a potential waterworld, known as Kepler-62f, over a thousand light years from Earth; and the slightly further away, Kepler-442b, a rocky Earth-sized planet. All are considered potential habitable candidates. Closer to home, Gliese 1214b, which is just 42 light years away, is a super-Earth discovered in 2009 that could be hosting an ocean. Recently, Wolf 1061c, a super-Earth orbiting in the habitable zone of a red dwarf only 13.8 light years away, has been added to the list and is the closest potentially habitable planet discovered so far.

The potential of these exoplanets to support life is ranked using an index that includes their distance from the centre of the habitable, or 'Goldilocks', zone; their similarity to some of Earth's parameters; whether they would be suitable for vegetation; their potential for hanging on to a habitable atmosphere; their temperature and mass, where temperature is also being used to infer what type of life could survive. Obviously, these parameters are being computed for life 'as we know it', despite the fact that there may be many other kinds of biochemistry in the Universe. The criteria we are using might be missing things we cannot yet conceptualise, but they do give us a conservative estimate for the number of planets that could harbour life.

Luckily, the abundance of data we are currently collecting from the Kepler mission and ground-based telescopes will soon allow us to expand our gallery of potentially habitable environments and biochemistries. How do we do this? The exoplanets mentioned above are so distant that sending probes to them would take many thousands of years. So, for now, our observations are still made through remote detection from telescopes, data analysis and modelling. We've developed many techniques for discovering new planetary systems through remote detection, including transit photometry and timing, radial velocity, reflected light variations, Doppler techniques, pulsar timing and frequency, gravitational microlensing, and now, direct imaging through extremely powerful telescopes! The more data we have, the more accurate – and better – models we will be able to generate to characterise habitable alien worlds and biospheres. We can also use what we are learning from the exploration of our own solar system, and that of terrestrial extreme environments. From those, we have learned that some planets and moons can still support habitable environments without being in the habitable zone ... and that life, once started, is very resilient and is found everywhere. We are soon also going to probe the atmosphere of some of these exoplanets using powerful spectroscopes on telescopes. They will tell us about the gas composition of these planets, which can give precious clues about whether life could be present or not.

In addition to the planets themselves, the exploration of the solar system shows that moons also have potential as habitable environments, both within and outside the habitable zone – and there are many more moons than planets. Therefore, the discovery of exoplanets is not just about their own potential for hosting life, but also about how many potential habitable worlds might be orbiting around them. No exomoon has been detected yet, but it is just a matter of time.

What's next?

Fifty years of planetary exploration has transformed our understanding of planetary habitability and the range of conditions we consider habitable. In the past 25 years, and accelerating exponentially since the Kepler mission was launched in 2009, the discovery of an abundance of exoplanets has revolutionised our concept of how many habitable worlds could exist in a very small fraction of our Galaxy alone. Astronomy and astrophysics are also opening wide the potential for habitable worlds in the 100 billion galaxies now estimated in our Universe. The idea that we could be alone is simply completely at odds with statistics.

The life out there might be somewhat familiar to us or completely foreign. However, if our planet can be a representative sample of other potentially inhabited planets and moons, it seems that nature produces a lot more of the simple life than complex organisms. Furthermore, the Earth remained populated by microorganisms alone for over 2.5 billion years. The sum of circumstances it took for complex life to evolve might suggest that a substantial fraction of inhabited worlds are actually populated by simple life, and it could be that our solar system is a representative sample of the ratio of simple to complex life in the Universe.

Technology and instrumentation are advancing at a fast pace too. New and more powerful missions will soon continue the search for exoplanets, from the ground and from orbit. In the solar system, ExoMars and Mars 2020 will search for traces of early life on Mars in the coming years, followed a few years later by a mission to Europa that will assess the habitability and biosignature potential of this icy moon of Jupiter. How we go from demonstrating habitability to truly ascertaining the presence of life will depend on two key factors: one is of course the presence of life, and the other is our ability to recognise its signatures. In this regard, Mars could provide a good training ground, as its early environment was fairly similar to Earth's: the Mars missions have shown us that the bricks of life were present – and as I mentioned earlier, there is the real possibility that the two planets shared material early in their history.

However, celestial mechanics did not favour an early exchange of material between the Earth, Mars, and the icy worlds of the outer solar system. If life evolved on those worlds, it will be most likely something different that might not be easy to recognise. On the other hand, exotic physicochemical conditions in the outer solar system could be what ultimately helps us begin the intellectual transition from the concept of life as we know it to that of different biochemistries and metabolisms that will be commonplace beyond the solar system.

Ultimately, this quest is also about finding others who have made the journey towards advanced civilisations like us, and who we might be able to contact one day. Our techniques are primitive and our exploration toolbox still limited. Today, radio and optical astronomy remain the basic tools in the search for extraterrestrial intelligence. We need to expand our approach and use our imagination – deploying broader scientific nets, which may even include greater understanding of interspecies communication here on Earth, and that of life's interaction with the environment and the surrounding Universe, and much more. We should not be afraid to push intellectual boundaries and bring a multidisciplinary approach to this quest. The baby steps we are taking today in planetary and space exploration will be remembered as those that took us to that encounter with others who also made that journey, far away from the Earth. When this contact might take place, nobody knows. However, what matters is that we have now started on that voyage.

17

Are They Out There? Technology, the Drake Equation, and Looking for Life on Other Worlds

Sara Seager

Somewhere far away, a living, breathing world peacefully orbits its star. Inhabiting the world is a dynamic ecosystem full of thriving simple bacteria. The life itself has no consciousness or intelligence, but the planet as a whole is an active world, connected through cycles of geophysics, chemistry and biology in a landscape with liquid water oceans, continents, mountains and volcanoes. We imagine there could be millions or even billions of such planets in our Galaxy.

Why would I, an astronomer, possibly speculate on the existence of life in the Galaxy? For three reasons. First, we now know that small planets are common. Second, water, a requirement for all life as we know it, is common. Third, the ingredients for life appear to form easily.

Our Galaxy is teeming with planets. Astronomers have found thousands of planets and planet candidates from several different planet-finding techniques. Furthermore, there is compelling evidence that all stars have planetary systems. In addition, observations of very young stars of all kinds show discs of leftover dust and gas that is expected to form planets. The pioneering Kepler space telescope – launched in 2009 and still continuing past its prime mission – has enabled the discovery of thousands of small rocky planets and planet candidates. As many as one in five or one in ten stars like the Sun could have an approximately Earth-sized planet in a favourable orbit such that, as heated by the star, the planet's surface is not too hot, not too cold, but just right for life.

Water is a very common planetary building block, and some scientists postulate that all terrestrial planets should be born with water. Water is locked away in minerals, delivered by what are known as planetesimals: small objects floating in space made up of rock, dust, ice and other materials that come together to help build the rocky planets. Crucially, they do so with such energetic collisions that any trapped water can escape. After a violently energetic era of formation, the planet cools and the water vapour can form a water ocean. Water is also delivered to a planet in the form of ice, by comets or asteroids. Some planets may lose water oceans, if they are too hot or too close to a star with powerful stellar winds, but on the whole there should be many planets with liquid water.

We see the ingredients for life in so many environments by way of organic molecules, the building blocks of life. Using ground-based radio telescopes, astronomers have observed large organic molecules in cold gas clouds deep in the interstellar medium. Amino acids, a class of biologically important molecules that play a central role in life on Earth, in proteins and as intermediaries in metabolism, have been found in a variety of carbon-rich meteorites. Even the very cold Titan, one of Saturn's moons, has molecules made of the elements that are required for life as we know it.

Some biologists have angrily disputed my statement that our Galaxy could be filled with planets containing life. After all, we do not understand how life originated on Earth, so how can we possibly have any confidence in anticipating life's existence elsewhere, much less life's ubiquity? Aside from the above three solid arguments, I must confess that I allow myself to speculate and daydream, because I am part of the first generation who has it within our reach to find signs of microbial life on exoplanets by searching for the gases they produce using sophisticated next-generation space telescopes. On a daily basis, my conviction that we must carry the search for life forward is growing.

The James Webb Space Telescope

In the fall of 2018, at the European spaceport in Guyana, the NASA/ESA James Webb Space Telescope (JWST) will be folded up in an Ariane rocket, awaiting launch. The JWST will have completed an intense journey from its conception in the mid-1990s. On launch, JWST will rocket to space and within the first week several precisely timed deployments (to unfold the primary mirror, secondary mirror and sunshield elements) have to go just right. JWST will take about a month to travel a million miles through space, far away from the heat and light of Earth that contaminates astronomical measurements. The JWST is often called the next generation Hubble Space Telescope, as it has a much larger mirror area and works at infrared wavelengths.

With JWST, we have our first opportunity ever to search for signs of life on a few choice planets. We will target small rocky planets, and observe their atmospheres, looking for gases that are far out of chemical equilibrium with the rest of the atmosphere. Perhaps surprisingly, the most compelling example is oxygen, which fills Earth's atmosphere to 20 per cent by volume – but without plants or photosynthetic bacteria, Earth's atmosphere would have virtually no oxygen. Today, planetary scientists are heavily debating what are referred to as 'oxygen false positives': scenarios in which oxygen might be produced without help from life. Astronomers are also considering a wide range of other biosignature gases, including methane, nitrous oxide, dimethyl sulphide, and others. We won't know if the suspect gases are produced by tiny microbes, enormous animals or intelligent humanoids, or if life producing the gas is carbon-based or something more exotic. To begin with, we have to focus on what life does – it metabolises and produces by-product gases – not on what life is.

The method we will use to observe rocky planet atmospheres with JWST is one I invented in a paper published in 2000 – although I should stress that there are rarely true inventions, just ideas built on past work. The method is as follows. Some planets pass in front of their stars ('transit') as seen from Earth. The planet's orbit has to be aligned just so, and only a small fraction of planets have this fortuitous alignment, because stars' rotation axes (and to some extent the planetary orbits) are randomly oriented with respect to our line of sight. As the planet transits the star, some of the starlight passes through the planet's atmosphere – but not all of it. At some wavelengths the planet's atmosphere absorbs starlight more than at other wavelengths. By carefully observing wavelength-by-wavelength, we can identify which gases and in some instances how much of them are present in the planet's atmosphere. There are many subtle caveats to the method, but, suffice to say, only 16 years after my prediction, hundreds of scholarly papers have been written on the topic, with many advances made by my team and others, and dozens of exoplanet atmosphere observations, most successfully with the Hubble's Wide Field Camera 3 instrument.

While the JWST can study individual planets, it cannot survey hundreds of thousands of stars needed to find the planets in the first place. Instead, we have the Transiting Exoplanet Survey Satellite (TESS), an MIT-led NASA mission which is specially designed to find a pool of small planets transiting small stars. TESS is a two-year survey of the entire sky, and has four identical, highly optimised, wide-field cameras that together can monitor a 24 degree by 90 degree strip of the sky. By monitoring each strip for 26 days and nights, TESS will tile the northern hemisphere sky in the first year and the southern hemisphere sky in the second year. It is scheduled for launch in fall 2017, aboard a SpaceX Falcon 9 rocket, and will go into a very eccentric, inclined orbit around the Earth. The TESS team will find a pool of fifty rocky planets for astronomers, a few of which will be in their star's Goldilocks zone: not too hot, not too cold, but just the right surface temperature for life. It is these precious few planets that we will propose to observe with the JWST to look for biosignature gases in the planet atmosphere. It's not an easy job: we will most likely need to observe many transits, each lasting for some hours.

The 'Seager equation'

How realistic is it that with the combination of TESS and JWST we will find signs of life on other worlds? Well, we have to get very, very lucky: but there is a chance, and we will be taking that chance. To illustrate, I use an updated version of the renowned Drake equation, devised by the American astronomer Frank Drake in 1961 to help us calculate the probability of intelligent life existing elsewhere in the Universe. Like the Drake equation, which was really intended as a summary of the main concepts that go into detecting signals from intelligent life in the Galaxy rather than a means to a definite answer, my revised equation is descriptive rather than predictive. It will show what we can quantify – what we *know* – and what must remain speculative.

Let's first take a look at the original Drake equation. This formula provides us with a rough value for the number, N, of communicative extraterrestrial civilisations in the Galaxy and is calculated by multiplying seven other numbers together:

$$N = R_* F_p n_e F_t F_i F_c L,$$

where R_* is the rate of star formation in the Milky Way (Drake assumed this to be 10 new stars per year); F_p is the fraction of stars with planetary systems (taken to be 0.5); n_e is the number of potentially life-hosting planets (with an 'ecosystem') per star (taken to be 2); F_t is the fraction of planets that develop life (taken to be 1);

F_i is the fraction of life-bearing planets with intelligent life (taken to be 0.5); F_c is the fraction of civilisations that have developed technology that releases detectable signals into space (taken to be 1); L is the length of time the civilisation is releasing such detectable signals (taken to be 10,000 years).

In the Drake equation, the first three factors, R_*, F_p and n_e, are measurable; the other four are not, and arguably never will be. Nevertheless, Drake arrived at the optimistic value for the number of communicative extraterrestrial civilisations in our Galaxy of 50,000.

At the time the Drake equation was devised, our main method of detecting extraterrestrial life was to listen for radio signals from alien civilisations. However, as we have seen, we now have more sophisticated means at our disposal: the search for life beyond the solar system by way of biosignature gases in exoplanet atmospheres has accelerated towards reality in recent years, warranting a new kind of descriptive equation.

In the fashion of the Drake equation, we may estimate the number, N, of planets with detectable signs of life by way of biosignature gases:

$$N = N_*F_Q F_{HZ} F_O F_L F_S,$$

where N_* is the number of stars in the survey; F_Q is the fraction of those ('quiet') stars in the survey that are suitable for planet-finding; F_{HZ} is the fraction of this subset of stars that have rocky planets in the habitable zone; F_O is the fraction of these planets that can be observed subject to our current limitations; F_L is the fraction of those planets that have life; F_S is the fraction of those planets with life that produces a detectable biosignature gas by way of a spectral signature.

Since I first presented my equation at a symposium in honour of Dave Latham's contributions to exoplanet science, Exoplanets in the Post-Kepler Era Symposium, held in Cambridge, MA, May 2013, the equation has been dubbed the 'Seager equation'. A more scholarly treatment of the equation than what follows is also available.

Let's go through the numbers and focus on the first four that we can at least say something about. The suitable number of stars in the TESS all-sky survey, out of the millions observed, is estimated, from galactic models, to be about $N_* = 30,000$. Out of these stars, about 60 per cent are expected to be quiet enough (that is, not highly variable in brightness) to allow for the detection of small planets in orbit around them. This means that $F_Q = 0.6$. This number is approximate.

Next, the fraction of rocky planets in the habitable zone is about 24 per cent (based on Kepler data) – that is, $F_{HZ} = 0.24$. The constraints due to the orientation of the orbits of these planets about their stars limits the fraction of stars with planets that can be observed by TESS because they pass (transit) in front of their stars to be about 10 per cent. However, only about 1 per cent of these will have stars bright enough for a planet's atmosphere to be observed in detail. In other words, we have $F_O = 0.1 \times 0.01 = 0.001$.

If we multiply these first four numbers together we have $N_* F_Q F_{HZ} F_O \approx 4$, which means we just have two more quantities that we need to assign values to (F_L and F_S). To keep things simple, we can also summarise the discussion so far as

$$N \approx 4 F_L F_S.$$

Very detailed calculations of the TESS planet yield – essentially the product of the first four terms of the Seager equation – are provided by a computational simulation where a value of between 2 and 7 is obtained.

We now turn to the discussion involving the terms F_L and F_S. All we can do for now is guess at the fraction of habitable planets that have life. Let us be optimistic and say that half of all potentially habitable planets do indeed have life: $F_L = 0.5$. Then does life generate gases that can accumulate in a planet's atmosphere and are spectroscopically detectable? Let us speculate that $F_S \approx 0.5$. These values, multiplied together, yield the number of planets with detectable signs of life from the TESS/JWST combination: $N \approx 4 \times 0.5 \times 0.5 = 1$.

Clearly I have tweaked the numbers to yield this favourable yet weak result: TESS and JWST may enable us to detect the presence of life in our little corner of the Milky Way Galaxy. This is my way of stating that we have to get very, very lucky to observe biosignature gases in the coming decade. We can, however, repeat emphatically that the TESS/JWST combination is our first chance in human history to search for signs of life by way of biosignature gases on rocky exoplanets.

What of the future?

What if the JWST does not find any signs of life in the atmosphere of any small rocky exoplanets? Astronomers are working hard to make the next generation of telescopes a reality, using a technique called direct imaging. This differs from the traditional transit technique of observing an exoplanet by measuring the light from its star as the planet passes in front of it. Instead, they aim to capture the much fainter light from the planet itself. There are already lab demonstrations showing that highly sophisticated light-blocking techniques required for direct imaging can work. I recently led a team on one of the many direct imaging concepts, called the Starshade, a giant specially shaped screen tens of metres in diameter that would fly in formation tens of thousands of kilometres from a space telescope, blocking out the starlight to a stunning 1 part in 10 billion so that only planet light enters the telescope. Of course, planets do not produce their own light and only shine because of reflected starlight. This technique enables us to search for signs of life in exoplanet atmospheres in a different way from the transiting method for which JWST is suited. We could, in principle, use the Seager equation for the Starshade survey as well.

I have full confidence that my generation of astronomers have the tools, know-how and strategies to find biosignature gases – *if* they are out there. But if life, or life that produces spectro-scopically active gases, is rare, then we will probably be out of

luck. Therefore, if the TESS/JWST or Starshade (or an equivalent mission using another direct imaging technique) do not find
biosignature gases, or even tentative signals, then we will have to
pass on the quest to future generations. Currently we know how to
build a space telescope out to about a 12-metre-diameter aperture
or even larger – perhaps up to 15 or even 20 metres. Beyond that,
the next generation will need to conceive of and implement a new
paradigm for space telescopes involving entirely new techniques.
Perhaps this involves the space-based construction of telescopes,
using technologies we can barely conceive of today.

A complication is that we can never be completely certain
that we have found signs of life on another planet just from gases
produced by life. The reason is what is known as a false positive
(potential indicators of life that could also be produced through
other means). Astronomers and planetary scientists are concoct-
ing numerous schemes under which a gas could be produced by
means other than life, such as in volcanoes or via various chemical
reactions in the atmosphere, or with the planet's rocks or oceans.
While astronomers are working hard to find the atmospheric
context in which a gas may or may not be a false positive, the observations of a planetary atmosphere across a wide wavelength range
and with sufficiently high spectral resolution may not be possible in
the near future. We could, in some cases, be highly confident, and
in others less so, perhaps ending with a statement to the world that
we have found suggestive evidence for life, but are only *somewhat*
confident (i.e., assigning a probability to the identification).

Despite a long road ahead, astronomers and astrobiologists
are confident that the search for, and detection of, biosignature
gases is within reach. The coming decades offer hope of extensive
progress in finding and characterising other Earths, and with it, of
course, the hope of detecting the all-important biosignature gases.

18

Good Atmosphere: Identifying the Signs of Life on Distant Worlds

Giovanna Tinetti

The Earth's atmosphere

On 14 April 1969, the Nimbus 3 probe was launched to monitor the Earth's atmosphere from an orbital altitude of about a thousand kilometres. Among the instruments on board was a spectrometer called IRIS, an instrument that analyses the spectrum of light it receives and from which it is able to identify the various chemical elements and compounds that the light has passed through.

The IRIS spectrum reveals, for example, the presence of water vapour (H_2O), carbon dioxide (CO_2) and ozone (O_3) in the Earth's atmosphere. In fact, because these molecules show up as a unique pattern of spectral lines in the infrared region, a part of the electromagnetic spectrum just beyond the visible, it is relatively easy to identify their presence. By contrast, other molecules such as nitrogen (N_2, around 78 per cent of Earth's atmosphere) and oxygen (O_2, 21 per cent) are not detectable in the IRIS spectrum because neither of these molecules shows a distinctive pattern in the infrared. Of course, the Earth's atmospheric composition was known well before the Nimbus 3 measurements, but that was the first time we had looked down from space at the light coming up from the Earth's atmosphere in the infrared. Today, however, examination of the atmospheres of remote planets far from our own solar system is one of the most promising avenues of investigation as we search for extraterrestrial life – namely, if we cannot yet detect alien life directly, what are the chemical clues that life might exist on distant worlds?

In essence, the question I wish to address in this chapter is this: how can a planet's atmosphere reveal the potential presence of life? To answer this question, we should first think about how our own atmosphere came to be the way it is, for it has not always been as it is today. At the time of the formation of our planet, about 4.5 billion years ago, it was probably composed mainly of hydrogen and helium, the most abundant gases in the proto-planetary disc where planets formed. Most likely, this primitive atmosphere did not last long: hydrogen and helium are very light gases and it is difficult for a planet as relatively small as the Earth to retain them gravitationally. In addition, these gases would have been stripped away by the solar wind, the stream of high-energy particles emitted by the Sun, which was probably more intense in that remote epoch, while the proto-Sun was still in the early stage of its evolution. The frequent collisions with other celestial bodies, such as asteroids or planetesimals, also contributed to the loss of the primordial gaseous envelope that surrounded our planet.

Our atmosphere was then modified, probably by a combination of repeated volcanic eruptions – which were commonplace in an era in which the interior of the Earth was still very hot and in the process of cooling – and impacts with comets and asteroids. During a volcanic eruption, large amounts of water vapour, carbon dioxide and sulphur products are emitted into the atmosphere, which is why it is not surprising to find these molecules in the Earth's atmosphere today. It is currently thought that nitrogen and most of the water were brought to Earth by asteroids. Since carbon dioxide, molecular nitrogen and water vapour are heavier than hydrogen and helium, this atmosphere has been longer lasting than the early hydrogen and helium one. One reason for this is the Earth's magnetosphere, a magnetic field that surrounds the Earth and its atmosphere, protecting it and slowing the erosion caused by the solar wind. The successful combination of temperature and magnetic field means that the Earth not only has an atmosphere but also a water cycle. On Earth, water that has evaporated from the surface and from the oceans largely condenses at a certain altitude, forming clouds that then discharge the condensed water vapour in the form of precipitation. By contrast, Venus's atmosphere is so hot that the evaporated water can't condense into clouds. This, combined with Venus's lack of a protective magnetosphere, has caused an irreversible loss of water from that planet over millions of years. Today Venus is an extremely dry planet because of these processes.

All this explains the origin of the gases in the Earth's atmosphere, with two exceptions: molecular oxygen and ozone, whose current concentrations are far from negligible. Molecular oxygen, in particular, is a very reactive molecule that easily reacts chemically with the other compounds. It is therefore difficult to explain, from a chemical point of view, why just over a fifth of the Earth's atmosphere is molecular oxygen.

Unless…

The signature of life

On Earth, molecular oxygen (O_2), made of two oxygen atoms bonded together, and, as I'll explain below, ozone, is produced by living things. Through photosynthesis, terrestrial higher plants such as trees and flowering plants are an (almost) inexhaustible source of oxygen, which explains its abundance in our atmosphere today. Ozone is a molecule consisting of three oxygen atoms (O_3), created when molecular oxygen splits and recombines: so ozone is considered an indication of the concentration of oxygen on Earth. Before the appearance of life on Earth, the amount of oxygen was negligible, as revealed by the chemical composition of some very ancient rocks. Today we believe that the first living organisms on our planet – called prokaryotes – appeared around 3.8 billion years ago. They were relatively simple organisms and the progenitors of today's bacteria. Simple but very tenacious and entrepreneurial: for another 600 million years, the prokaryotes colonised the Earth, living happily and undisturbed, testing all possible combinations of 'food' and metabolisms. For example, the *methanogens*, as the name suggests, generate methane as waste product. Another example is the protobacterium *Shewanella putrefaciens* that feeds on ferric iron (Fe^{3+}) and expels ferrous iron (Fe^{2+}), storing the chemical energy this produces. Other prokaryotes fed on sulphates, nitrates and cyanides. If we could record an infrared spectrum of the Earth at that time, we would observe the signature of water vapour, carbon dioxide, and perhaps a bit of methane and nitrogen or sulphur compounds, outgassed by these organisms. We would certainly not have seen the ozone signature. For most of our early ancestors, oxygen was actually the equivalent of hydrogen cyanide to us – that is, a lethal poison. It was only when the concentration of oxygen increased to high enough levels in the atmosphere that Darwinian selection prevailed and some prokaryotes learned to make use of oxygen. That was the key to success.

The evolution from prokaryotes to more advanced single-celled organisms (eukaryotes), and then from eukaryotes to multi-cellular organisms, lasted about a billion years, and was one of the most important events in the history of life on Earth. And it is the use of oxygen, which releases a far greater amount of energy than anything used by previous life forms, that is the key to understanding the complexity of life on Earth. These complex processes that life adapted to use have assumed different forms over the course of hundreds of millions of years, taking evolutionary paths that suited the environment of the time. Over the last 500 million years in particular, with a moderate climate and plentiful food, gigantism was a successful survival strategy, as was seen during the age of the dinosaurs. But at a time of food shortage and climatic difficulties, warm-blooded and adaptable smaller animals prevailed – the mammals. Our progenitor prokaryotes did not disappear completely from the face of the Earth, however, but had to move away from oxygen to hidden corners, such as geothermal springs and rocks with high concentrations of metals and silicates, which even today are home to exotic microbial communities, probably very similar to those that populated the Earth in a primordial age.

The other stroke of genius of life on Earth was the evolution of the most important biochemical process in the known Universe: photosynthesis. This is the process that allows plants and certain bacteria to store the energy from sunlight in chemical bonds within living cells. Without photosynthesis, complex life could not have developed for lack of food or renewable energy to ensure the survival of complex organisms. Higher plants have a photosynthetic pigment, chlorophyll, capable of capturing solar light and, eventually, producing glucose and molecular oxygen. Therefore, the presence of high levels of oxygen in an atmosphere is a good example of a 'biosignature': something that provides evidence that life exists, or may have existed in the past, on a planet. For the sake of completeness, I should mention that sulphur purple bacteria are very ancient photosynthetic organisms that use hydrogen sulphide (H_2S) rather than water in the process of photosynthesis. These bacteria, unlike our green houseplants, do not produce oxygen.

Lovelock and the definition of biosignature

Spectroscopic data from satellites such as the Nimbus 3 has profoundly changed our perception of life on Earth. Viewed from the outside, life on Earth appears as one of many possible forms of life, detectable, at least in principle, by pointing a telescope at another planet. The Venera and Mariner space missions, not finding traces of life on Mars and Venus, marked the end of the myth of habitability for these two planets. For Mars, there is still the hope that some forms of underground life or trace fossils of extinct life might be found (as was discussed by Monica Grady in Chapter 7). In the solar system, there is still a glimmer of hope that life has formed on some of Jupiter's or Saturn's moons, thanks to the energy generated locally from tidal phenomena caused by those giant planets' gravitational forces. However, we have to resign ourselves to the idea that the only form of complex life in the solar system resides on Earth.

Now, thanks to developments in technology, we are no longer limited to our own solar system in our search for life. The hope of finding life elsewhere has increased in proportion to the number of newly discovered exoplanets (around two thousand as of early 2016). We have only very basic knowledge about most of the known exoplanets, such as their mass and rough size. But very recently we have learned how to observe the chemical composition of their atmospheres and measure the planetary thermal structure. This has become possible through two techniques that allow us to probe the atmospheres of exoplanets: transit and eclipse spectroscopy, and direct imaging spectroscopy. Transit and eclipse techniques rely on differential measurements of the star and planet system, due to changes of the planet's position relative to the star it orbits, that is, when the planet passes in front of the star or goes behind it. Direct imaging spectroscopy is a new and remarkable technique and was discussed in the previous chapter by Sara Seager.

Thanks to the Hubble and Spitzer telescopes, and ground-based observations, we have started to probe the key chemical components and thermal properties of the most promising transiting exoplanets. These are mainly hot gaseous planets orbiting very close to bright stars. Most recently we were able to push the instruments and data analysis techniques to capture the main atmospheric features of planets called super-Earths – solid planets with masses below ten Earth masses – though the super-Earths analysed thus far are still much too hot to be considered interesting targets for habitability.

Recently, direct imaging techniques have started to provide new insight into the atmospheres of young gaseous planets orbiting at large distances from their parent star. The most significant current projects using this technique are the Gemini Planet Imager at the Gemini South telescopes in Chile and the SPHERE (Spectro-Polarimetric High-Contrast Exoplanet Research) project at the VLT (Very Large Telescope) in the Atacama desert, also in Chile. Other notable instruments for direct imaging include projects attached to telescopes in Califonia and Hawaii.

How, then, can we know whether a planet is habitable, or maybe even inhabited? Obviously the chemistry and the state of their atmospheres is critical to understanding the origin and evolution of these planets, and is vital if we are going to formulate any hypothesis of habitability. Scientists have been racking their brains over the last fifty years on this issue, and we may well have some answers in the coming decades, although many obstacles still stand in our way. For instance, although the laws of physics are universal – the same in London, on the Moon, and on Proxima Centauri – and, on a large scale, the Universe seems to be homogenous, we do not have a definition of life that can be generalised beyond what is found on Earth. For example, since on Earth oxygen and ozone are of biological origin, should we then be looking for these two molecules on other planets as proof of their habitability? That is, are these universal biosignatures, or do they appear just on Earth?

James Lovelock was among the first scientists to try and answer these questions in a rigorous manner. His pioneering articles on extraterrestrial life date back to the early sixties, in which he tried to tackle the issue of finding a general definition of life that is both scientific and pragmatic. His interest in the subject was stimulated by the then imminent launch of the NASA probes Viking 1 and 2, designed to land on Mars and to seek, among other things, traces of life on its surface. Lovelock was unconvinced by the various mechanical devices proposed by his colleagues to search for life on Mars, including small traps for possible Martian critters. Lovelock argued instead that it is Mars's very thin atmosphere that should be studied instead in order to understand if the red planet is inhabited or not. The atmosphere of a lifeless planet is – as Mars's atmosphere was found to be by Viking 1 and 2 – very close to chemical equilibrium, which led Lovelock to conclude that Mars was lifeless. As explained earlier, the concentration of oxygen and ozone in our atmosphere increased following the emergence of multicellular species, so the composition of the Earth's atmosphere today shows the unequivocal evidence of the presence of life that is discharging this oxygen into the atmosphere. If living things were to disappear from the face of the Earth, oxygen and ozone would rapidly disappear as they react with other chemicals until equilibrium is achieved. Other examples of terrestrial biosignatures are seasonal fluctuations in the concentration of CO_2, as vegetation flourishes in the summer and dies back over the winter. There is also something called the 'red-edge signal'. This is such a clever notion that it deserves explanation. During photosynthesis, plants absorb sunlight mainly from the visible part of the spectrum, but not the longer-wavelength infrared light, which they just reflect back. This 'reflective property' of foliage is easily recognisable from satellite measurements. When a graph is plotted of the intensity of light against wavelength, we see a sharp drop (the red edge) when we go from longer (infrared) wavelengths to shorter (visible light) wavelengths.

The idea of using the atmospheric composition of a planet to investigate how habitable it might be can be applied to exoplanets as well. Lovelock's definition of biosignature, essentially the chemical disequilibrium caused by the presence of living organisms on the planet, is currently the only scientifically rigorous one we have. But it is not perfect, and it is possible that inhabited worlds may not stand out from a multitude of similar planets when observed. In particular, we don't have a thorough understanding of the chemistry of exoplanet atmospheres – we don't know whether they are mostly in equilibrium or whether they are driven away from equilibrium by non-living processes, as has been suggested by computer simulations. All we can do for now is study and observe a very large number of planets in our Galaxy with different sizes, temperatures and host stars to understand what potential alien worlds might look like. Without this information, and hence the ability to understand the big picture, it would be somewhat speculative to identify a planet as inhabited based solely on the definition of biosignature given above.

The obsession of the Earth's twin

In its early days, the search for planets outside the solar system was strongly inspired and influenced by the search for an Earth twin, a planet that would exactly resemble ours. But the search for a mirror image of our planet, in our Galaxy or in the wider Universe, is not only unscientific, it is not even particularly interesting. The thought that the Earth is the only, or the most interesting, model of habitable planet appears to be a mix of arrogance, parochialism and anthropomorphism, seeing ourselves and our world as the centre of the Universe, as we did before Galileo. There is nothing particularly special about Earth. In fact, the statistics of the solid planets we know about today is a warning to broaden our horizons.

NASA's Kepler satellite was conceived more than twenty years ago to find Earth twins around Sun-like stars. In the statistical analysis of the Kepler data, the size of the Earth does not seem to be so crucial, but rather an accident in the spectrum of possible sizes for solid planets. By this I mean that planets about twice or half the size of the Earth should be, in principle, equally habitable. And what about our Sun? After all, we know it is a fairly standard star, not too large and not too small, in the middle of its journey through life. Could a habitable planet evolve around smaller and colder stars than the Sun, or larger and hotter ones? I do not see why not. Leaving out the extreme cases – stars too massive and unstable or too active – there is still a wide spectrum of possibilities.

So, what else can we measure in our search for habitable worlds? Well, temperature is one property: if we think that life must be based on carbon chemistry, and chemical bonds similar to the ones developed by terrestrial life, then the planetary temperature cannot be too different from Earth's. This sounds at odds with my resounding claims of wanting to broaden the horizons of possibilities, but it is true that life on Earth is based on the most common elements in the Universe: hydrogen, carbon, nitrogen, oxygen. In addition, a long list of complex organic molecules has been identified in stellar or planetary formation regions, or comets. It includes amino acids, the building blocks of proteins, and precursors of nucleotides, the components of DNA and RNA – our genetic material. We are certainly not made of unique or rare chemical compounds, and it is precisely because the components of life as we know it seem to be ubiquitous in the Universe that it is logical to adopt carbon-based life as a working hypothesis. It then follows that planetary temperature cannot be random: temperatures that are too high can cause irreversible damage to the structures of organic molecules. Similarly, temperatures that are too low would slow molecular reactions, making it harder for life to get going in the first place.

Life on Earth is also strongly dependent on water as a chemical solvent. It has long been debated whether other solvents, for example ammonia, could supply the chemical functions of water, but this remains only a hypothesis. Therefore, if we want to proceed with a rigorous scientific method, we cannot rule out liquid water from the list of ingredients necessary to life. After all, we know that liquid water is crucial to the health of most complex organic molecules, and this need for water in its liquid state puts a limit on the range of temperatures and pressures on host planets.

Now that we have broadened our search for life outside of our own solar system, we are looking at worlds so far away that analysis of their atmospheres may be one of the only methods available to us to determine whether they are habitable or not. What constitutes a biosignature and how we can explain disequilibrium in an atmosphere are questions of vital importance in this quest. And, impossible though it is to be certain, if we find a planet whose atmosphere changes over time and contains a high proportion of water vapour and oxygen … well, who knows?

According to the Habitable Exoplanets Catalog (http://phl. upr.edu/projects/habitable-exoplanets-catalog), there are roughly 33 habitable planet candidates – that is, solid planets with the temperature range needed for liquid water.

19

What Next? The Future of the Search for Extraterrestrial Intelligence

Seth Shostak

Are there beings that inhabit the realms beyond Earth? Many early cultures assumed the answer was yes, and their myths described a sky teeming with gods and creatures.

However, when telescopes showed that the peripatetic objects known as planets are worlds unto themselves, aliens became less like Greek gods, and more like us. Yes, the heavens were inhabited, but usually by human analogues. This was true of both scientific speculation and, from the nineteenth century onward, science fiction. Our hypotheses about celestial beings seldom envisioned alien microbes, for example, despite the fact that these are likely to be the most common form of cosmic biology.

It's therefore hardly surprising that, in the popular mind, looking for life elsewhere is synonymous with seeking our peers. Inevitably, our efforts to do so presume that aliens are like us in important ways, or at least similar to the prognostications we make about our descendants. We combine these largely unspoken assumptions with our knowledge of physics and astronomy to fashion experiments such as SETI (the search for extraterrestrial intelligence) – attempts to detect electromagnetic signals that a faraway society might be deliberately or inadvertently lofting into space.

While the first SETI experiments were simple efforts to use sensitive radio astronomy equipment to listen for transmissions from nearby star systems, these experiments have been augmented with other efforts to look for either brief pulses of laser light or steady monochromatic light sources. These latter endeavours are known as 'optical SETI'. It should be noted that SETI is simply a generic acronym for experiments intended to prove the existence of extraterrestrial intelligence by looking for artificially produced electromagnetic radiation.

While seemingly casting a wide net, modern radio SETI searches have limited scope. In order to design practical experiments, researchers must make assumptions about frequencies and bandwidths, as well as the strength and duration of signals. Underpinning these technical concerns is the premise that our knowledge of physics is sufficiently complete for us to infer the preferred method for interstellar signalling. Unlike scientists of the nineteenth century who looked for flashes of sunlight from mirror-wielding Martians, we now reckon we can guess how any sophisticated society will communicate.

An even greater supposition made by those doing SETI research is that our galactic peers (or more likely, our superiors) have some motivation to project their presence. It is perhaps as support for this requirement that we speak of alien 'societies' – large, diverse cultures with a need for communication that would ensure copious signalling.

How SETI is done

The scheme used by the majority of SETI experiments today can trace its ancestry to an experiment known as Project Ozma, conceived and carried out by astronomer Frank Drake in 1960. The approach is to use large antennas, also known as radio telescopes, to scan the sky for narrow-band radio signals (or signal components) that would be the hallmark of a transmitter. This restriction to narrow-band signals may be seen as antiquated, given the increasing use of spread-spectrum transmissions on Earth. However, it's a limitation imposed by current technology, and may be ameliorated in the future as computing hardware improves. It's also the case that narrow-band signals promise the greatest signal-to-noise for any given power level. In other words, by putting as much of the transmitter power as possible into the narrowest range of frequencies, the resulting signal will stand out against the inevitable background noise of both the cosmos and the receiving apparatus.

SETI's antennas are either aimed at nearby star systems – so-called targeted searches – or are used to survey large tracts of the heavens by tessellating the sky into beam-sized areas. Note that either strategy has a very low duty cycle, or fraction of time spent observing in any given direction, and this implies that discovery of a signal will only occur if that signal is persistent.

This temporal limitation is also true of so-called optical SETI experiments. These use conventional mirror-and-lens telescopes to search individual star systems for brief flashes of light from distant solar systems.

While no confirmed signal proving the existence of extraterrestrial intelligence has ever been found, this failure is tempered by the fact that the number of star systems sampled is still quite small. Only a few thousand such systems have been reconnoitered with high-sensitivity equipment over a wide range of frequencies. So a reasonable question to ask is: How many stars need we examine before we have a good chance of success? Estimates of the number of galactic locations that might be transmitting signals of a strength that could be detected with today's SETI instruments range from 10,000 to 1,000,000. If the 'true' number of detectable transmission sites really is within this range, then finding a signal will likely require the scrutiny of several million star systems. While that's an interesting result, and encourages SETI's efforts, it's important to note that the estimates on which it's based are nothing more than opinion.

A fundamental assumption that could be wrong

Our mental image of the extraterrestrial intelligence we seek is a vague extrapolation of ourselves: we assume that whatever is out there will be roughly equivalent to us in technical sophistication. But the assumption that we are most likely to pick up a signal from a species functionally similar to our own is challenged by the following timescale argument:

1. Any detectible intelligence must have reached a technological level at least equal to our own. Indeed, it's virtually a requirement that its level be *greater* than ours, since SETI, as practised today, could not detect most terrestrial transmissions even at the distances of the nearest stars. To hear them requires that they have more powerful transmitters than we currently do.

2. The probability of success – as is often reckoned by the well-known Drake equation (see Chapter 17) – is encouragingly large only for transmitting entities that are long-lived, i.e., that are 'on the air' for extended periods of time. In this context, 'long-lived' is generally taken to be at least 5000 or 10,000 years. This means that the population of technically capable beings we are most likely to detect are members of a civilisation that is at least thousands of years older than our own.

3. If one believes the claims made by those working in the field, humans will invent generalised artificial intelligence (GAI) within this century. Even if this expectation is overly optimistic – even if the interval between inventing radio and inventing thinking machinery is many centuries – there seems to be no escape from the conclusion that the majority of the societies we have traditionally postulated as targets for SETI have in all likelihood

already managed to construct artificial intelligence that surpasses their own.

The conclusion from the above is straightforward: Since the development of GAI quickly follows the invention of radio and lasers, the bulk of the intelligence in the cosmos (as defined by an ability to communicate) is very likely to be machines, not living creatures.

This fact goes to the very heart of the assumptions made by SETI practitioners. In particular, it challenges the idea that we should focus our search on looking for signals that are either deliberately or accidentally launched from a biologically friendly world around another star. Biology may be the exclusive packaging for intelligence of the past and the present. But not of the future.

So how does this affect our SETI experiments? Machine intelligence requires an energy source and the raw materials for fashioning new and replacement parts. In order to grow – that is, in order to increase its capability for computation – greater amounts of both are needed. While planets provide metals that we might reasonably assume are the building blocks used in the construction of these synthetic sentients, such worlds are not the only possibility. Asteroids have a far higher percentage of many metals than Earth's crust, a circumstance that will likely be true of other solar systems. As far as energy is concerned, the radiative flux available on a planet is limited by how much energy is provided by its own sun (in the case of Earth, approximately 10^{17} watts). That is a hard constraint, but one that can be easily circumvented by moving out into space and away from the alien civilisation's home planet.

These considerations challenge the wisdom of SETI searches that target planets. In particular, if we generalise from our own predicted technological trajectory, then the time between the development of equipment able to generate strong radio or light signals and the invention of GAI is mere centuries. If we combine this with the expectation that only long-lived transmitters have a reasonable chance of detection, we then see that the traditional assumption that intelligence will be more likely to be found on a biofriendly world could be wildly mistaken.

In particular, the excitement with which the SETI community greets every new discovery of a habitable planet might be misplaced. At the very least, looking for intelligence on biofriendly worlds is rooted in the assumption that the biological founders of GAI will continue to produce signals, for commerce among themselves or with other biological organisms, or possibly with departed GAI. In other words, if biological intelligence continues to exist after the invention of GAI, it may still be a reasonable target for a conventional SETI experiment. But it's also possible that synthetic intelligence could completely supplant its living ancestors.

What really happens to a society once synthetic intelligence is invented is quite obviously beyond our ken. There is no clear reason to expect that GAI devices would intentionally obliterate the living beings that first built them, any more than *Homo sapiens* seeks to intentionally destroy its ape-like ancestors. But even so, many of the simians that preceded us are threatened with extinction. In our own case, it's at least possible that once GAI establishes a presence on Earth, it may so dominate the planet's resources – material, energetic and geographic – that *Homo sapiens* will be marginal-ised in the way that great apes are. Societies of communicating biological intelligence could be very short-lived, and consequently the chances of finding them could be small.

What about finding GAI, then? As noted, it seems reason-able that the more accomplished GAI devices would eventually move away from their natal planet and into space, in search of the massive energy and material resources that would allow them to grow. How far they would emigrate is unclear, but smaller stars (such as the Sun), while seemingly preferable for spawning biology, are not exceptionally powerful energy sources, and consequently not the most attractive locales for post-human entities eager to situate themselves near sources of plentiful energy. O-type stars, which are far larger and more luminous than stars like the Sun, have luminosities that are a million times that of a solar-type star. These brighter lights constitute approximately 0.001 per cent of all stars, and are therefore sparse: they are separated, on average, by several hundred light years. SETI searches that examine late-type stars could easily accidentally miss some of these high-energy targets.

Where to look

If we can't confidently restrict the locales of the intelligence we seek to habitable planets, what other places might we search? Clearly until and unless we develop GAI ourselves, any speculation on its location and behaviour is highly uncertain. However, a few plausible suppositions about strategies would include the following:

1. *Search vicinities with high energy density.* An unbounded increase in knowledge and the ability to reason seem desirable for machine intelligence, if for no other reason than to forestall natural disasters or threatening competition from other devices. As noted, brighter stars are obvious candidates, as are the neighborhoods in which black holes are found, including galactic nuclei.

2. *Be sensitive to evidence of massive astroengineering.* Highly advanced intelligence may have constructed something we might trip across in our routine scrutiny of the skies. An obvious example would be to note unnatural infrared sources that might betray the presence of massive astroengineering projects. One of these might be a swarm of solar-energy-collecting satellites situated beyond a planet's orbit. Physicist Freeman Dyson proposed that advanced societies might construct such swarms to provide themselves with

a superabundance of energy. Dyson swarms around Sun-like stars are capable of generating ~10^{26} watts, far more than could be used on a planet without destruction of its climate. But orbiting GAI has no such limit on energy use, and consequently searches for the infrared signature of such Dyson swarms are, in fact, more attuned to finding synthetic intelligence than biological beings. However, this approach is largely at odds with the types of experiment generally pursued by SETI practitioners. In addition, such experiments can be done 'at home' using existing astronomical data sets.

3. *Look for signals travelling along likely communication corridors.* While it's unclear whether AGI would communicate with one another, it is hard to rule out the possibility that they would want to. Given the unlimited lifetimes of such intelligence, even long-distance communications might be feasible and of interest for exchanging information about the part of the Universe visible to one of these machines. The lines connecting black hole pairs are examples of possible communication corridors, as would be the lines connecting the nuclei of galaxies. One could look at any Galaxy in the anti-centre direction of our own, or black holes that have a cousin on the opposite side of the sky.

4. *Be sensitive to intermittent 'beacon' signals.* These might be sporadic or even systematic efforts by machines to locate other such devices, or disregarding the obvious hubris, perhaps even biological intelligence.

5. *Be alert to apparent violations of physics.* Given the capability and longevity of machine intelligences, they might have the ability to alter the cosmos at a fundamental level.

Some of the above require no more than being aware of the possibility of finding signs of non-natural phenomena in the course of conventional astronomical work. Others involve deliberate SETI searches, but not necessarily of the type conducted so far.

But what about the latter? SETI projects now underway – obedient to traditional SETI approaches – are greatly enlarging the number of star systems examined for signals. Would these have any chance of finding GAI?

They might. Today's radio SETI examines the sky with beam sizes ranging from about 0.5 to 15 arcminutes at frequencies near 1 GHz (and smaller at higher frequencies). Beam sizes are a measure of the tightness of focus of these instruments – the smaller the beam, the narrower the field of view of the radio telescope. The beam sizes in use correspond to patches on the sky that would completely cover a solar system, even one as close as 10 light years. Consequently, if the type of synthetic intelligence described here remains within its solar system of origin, then conventional radio SETI experiments have some chance of finding it (assuming it produces signals). However, optical SETI searches have, thanks to the laws of optics, much, much smaller beams, and if this type of experiment is used to aim at nearby stars, it might easily miss any GAI that has transported itself to very bright stars or other locales that are not obviously biofriendly.

These considerations suggest that sky surveys – experiments that look at as much of the sky as possible – rather than searches aimed at individual stars might be preferable strategies if the intelligence we seek is synthetic. In addition, there is a need for instrumentation that can reliably detect highly intermittent signals. The nearest extraterrestrial intelligence might plausibly be at a distance of 100 light years or more, which means that any radio signal sent between us and them would take 100 years to cover the distance, travelling at the speed of light. This means that whatever the intelligence is – biological or synthetic – it will not know of our existence, which has been betrayed by high-frequency, high-power radio signals only since the Second World War. They may possibly ping us with an occasional flash of radio or light, simply out of curiosity (and possibly motivated by the detection of biogenic gases in Earth's atmosphere, a signal of life on this planet that's been radiated into space for 2 billion years.) Frankly, the relentless transmissions to Earth that historical SETI experiments have assumed to be present are hardly guaranteed. Data processing set-ups that can confidently recognise intermittent signals are something any SETI effort would benefit from.

Aside from the possibility of accidental interception of signals occasioned by luck or by being situated along the communication pipe joining GAI extraterrestrials, is there any hope that sentient machines would deliberately signal in our direction? This too requires a highly speculative (and, some would say, doubtful) proposition: that they are motivated to do so. Humans communicate with their peers, and not with creatures that evolved long before themselves. Perhaps some signalling in the directions of biofriendly planets might take place, simply out of curiosity. Another possibility would be very broadly beamed transmissions, intended as 'hailing signals' to either locate or inform other GAI.

The consequences of success

The big-picture aspects of SETI considered in this chapter – and in particular its fundamental assumptions – may help us refine our experiments and possibly increase the chances of their success. But it would be irresponsible not to include some thought as to the ways in which a SETI success could bend the future arc of history.

The short-term consequences of a detection are likely to be no more dramatic than the initial reaction to Columbus's return to Spain after landing in the Americas. His discovery was ambiguous (had he, in fact, reached Japan?), and little was known of the culture that he encountered. In the case of SETI, the characteristics of the signal should tell us a few things about the location and physical status of the senders, assuming they're on a planet. If not, then we might not learn much – even as straightforward a parameter as the rough distance to the transmitter might be difficult to ascertain.

In other words, a SETI success would be exciting, but not because of the information we would garner. Rather, it would tell us that something intelligent is out there, a philosophically stunning result. It would be a big story.

In the long term, of course, it might be possible to tease out encoded information in whatever signal is found, and as noted earlier, this will almost certainly be from an intellect beyond our own. The consequences for religion, for our sense of self-worth, and for the future of our species can be only dimly perceived. It might be that we never really understand the signal we've found, and simply must deal with the implications of knowing that we're not unique. On the other hand, we might experience the kind of disruption that affected the Japanese when they were exposed to Europe's more advanced mathematics and science in the seventeenth century. We might lose confidence in our abilities to guide our own future, or even with our ability to try.

But even if we don't understand any message in a detected signal, it still might be possible to know that it comes from a non-biological source. This would underline the predictions made here that our future will include – indeed, will be dominated by – the development of generalised artificial intelligence.

Astronomers continue to track down habitable planets around other stars. It's likely that, within a year, as data from NASA's Kepler space telescope continue to be analysed, we will discover other worlds that are very much like our Earth. Such planets would be obvious candidates for incubating life, and possibly intelligent life. But the incubator is not necessarily where intelligence will stay. Indeed, given our own experience, it will likely leave the cradle shortly after inventing radio.

In other words, biological intelligence might be only a stepping stone to something far cleverer, something that is both longer-lived and more widespread than its protoplasmic precursors. There's a lesson in this: In our search for intelligence beyond the bounds of Earth, we should be careful not to be dinosaurs looking for other sauropods.

Further Reading

Pale Blue Dot: A Vision of the Human Future in Space, Carl Sagan, Ballantine Books 1997.

The Aliens Are Coming!: The Exciting and Extraordinary Science Behind Our Search for Life in the Universe, Ben Miller, Sphere 2016.

Extraterrestrial Civilizations, Isaac Asimov, Ballantine Books, New York 1980.

Alien Life Imagined, Mark Brake, Cambridge University Press, Cambridge 2012.

What Does a Martian Look Like? Jack Cohen and Ian Stewart, Ebury Press 2003.

Our Cosmic Habitat, Martin Rees, Princeton University Press 2011.

Life in the Universe: A Beginner's Guide, Lewis Dartnell, Oneworld 2007.

Life As We Don't Know It, Peter Ward, Viking 2005.

Shapes, *Branches* and *Flow* (trilogy), Philip Ball, Oxford University Press 2011.

The Second Law, Peter W. Atkins, Scientific American Library 1984.

The Biological Universe, D. J. Dick, Cambridge University Press 1999.

Goldilocks and the Water Bears: The Search for Life in the Universe, Louisa Preston, Bloomsbury 2016.

The Vital Question: Why Is Life the Way It Is?, Nick Lane, Profile 2015.

Life on the Edge: The Coming of Age of Quantum Biology, Jim Al-Khalili and Johnjoe McFadden, Black Swan 2015.

Are We Alone?, Paul Davies, Basic Books 1995.

Life Itself, Francis Crick, Simon & Schuster 1982.

The Goldilocks Enigma: Why is the Universe Just Right for Life?, Paul Davies, Penguin 2007.

The Eerie Silence, Paul Davies, Houghton Mifflin Harcourt 2010.

The UFO Book – Encycolopedia of the Extraterrestrial, Jerome Clarke, Visible Ink Press 1998.

UFO: The Government Files, Peter Brookesmith, Blandford Press 1996.

How UFOs Conquered the World: The History of a Modern Myth, D. Clarke, Aurum Press 2015.

Area 51 Viewers Guide, Glenn Campbell, Self-published 1993.

Area 51: The Dreamland Chronicles, David Darlington, Holt 1998.

Top Secret Majic, Stanton T. Friedman, Marlowe 1997.

Barlowe's Guide to Extra-terrestrials, Beth Meacham, Ian Summers, and Wayne D. Barlowe. Workman Publishing 1979.

Sharing the Universe, Seth Shostak, Berkeley Hills Books 1998.

Life Ascending: The Ten Great Inventions of Evolution, Nick Lane, Profile 2009.

Creation: The Origin of Life/The Future of Life, Adam Rutherford, Penguin 2013.

Abducted: How People Come to Believe They Were Kidnapped by Aliens, S. A. Clancy, Harvard University Press 2015.

Octopus, Jennifer Mather and Richard Anderson, Timber Press 2010.

The Soul of an Octopus, Sy Montgomery, Simon & Schuster 2015.

30 Second Brain, Anil Seth (editor), Ivy Press 2014.

*If the Universe Is Teeming with Aliens ... Where is Everybody?:
Fifty Solutions to the Fermi Paradox and the Problem of
Extraterrestrial Life*, Stephen Webb, Copernicus 2002.

Online

We've chosen not to include stand-alone web addresses, as they're
subject to constant change. As Dallas Campbell points out, there
is a vast amount of information about UFOs online. Some good,
some dreadful, some entertaining, some utterly delusional. Ian
Ridpath's site (currently www.ianridpath.com) is a good place
to start. Jenny Randles has a vast amount of experience in this
field (http://www.ufoevidence.org/researchers/detail40.htm), and,
of course, you should have a look at the *Fortean Times* (http://
subscribe.forteantimes.com), which has reported on the 'news of
the weird' with sense and good humour since 1973. In broader
terms, it's always worth looking out for alien-themes TED talks
(such as 'Why Mars might hold the secret to alien life' by Nathalie
Cabrol, March 2015; or Louisa Preston's TED-Ed original 'Why
extremophiles bode well for life beyond Earth'). You can find a
list of potentially habitable exoplanets at the Planetary Habitability Laboratory's site (http://phl.upr.edu/projects/habitable-
exoplanets-catalog). The Astrological Society of the Pacific's
'Universe in the Classroom' series provides clear tutorials on many
concepts to do with aliens and space (http://www.astrosociety.org/
publications/universe-in-the-classroom/).

Adam Rutherford's must-watch alien film list

Contact (1997): Jodie Foster decodes a regular repeating signal from the star system Vega, and works out that it's the instructions to build a wormhole to make contact with other civilisations. Carl Sagan wrote it, which explains why it works so well.

2001: A Space Odyssey (1968). There are few better films than this. The alien presence, in the form of a monolith is mathematically elegant, indicating that maths is a universal language: 1:4:9, the squares of the integers 1, 2 and 3.

The Thing (1982): a shape shifting alien tries to survive by imitating scientists in an Antarctic research station. Fun fact: the whole film is explained in the first lines that the dog is not a dog, but an alien. But it's said in Norwegian, so the crew are unaware that things are about to go south.

Alien/Aliens/Alien 3 (1979, 1986, 1997) The xenomorphs are humanoid in the first two, but this is because they have gestated inside people. In *Alien 3* (underrated in my opinion), it is canine, as it pupated inside a dog.

Attack the Block (2011). On Bonfire Night, aliens arrive in a south London council estate, and it's left to four teenagers and a nurse to defend civilisation from some wonderful extraterrestrials described quite accurately as 'big alien gorilla wolf motherf**kers'.

Plan 9 from Outer Space (1957): aliens arrive to enact their infamous Plan 9, which involves resurrecting the recently dead (primarily minor cult celebrities including Bela Lugosi, a wrestler and Vampira). Director Ed Wood's ultra-low-budget B movie is so bad it's good.

And some to avoid ...

Prometheus (2012): Life on Earth was created by some super
muscly humanoid aliens, who also left some clues to find
them; everyone dies through stupidity. Nothing about this
film makes any sense.

Signs (2002): Aliens who are allergic to water invade Earth,
a planet that is covered in water and harbours life that is
dependent on water. Not the best invasion plan ever.

Contributors

NATHALIE CABROL is an astrobiologist specialising in planetary science. In 2015, she was appointed head of the SETI Institute's Carl Sagan Center for the Study of Life in the Universe, and is also Principal Investigator of the SETI Institute NAI team, which aims to develop new biosignature detection and exploration strategies in support of the upcoming Mars 2020 mission. She is also an extreme diver and mountaineer, skills she has used in the course of her research.

DALLAS CAMPBELL is a television science broadcaster whose credits include some of the most high profile factual documentaries and programmes of recent years, for the BBC and beyond: *City in the Sky*, *Britain Beneath Your Feet*, *The Treasure Hunters*, *Supersized Earth*, *Airport Live*, *Stargazing Live*, *The Sky at Night*, *Egypt's Lost Cities*, *Bang Goes The Theory* and *The Gadget Show*. He also presented the BBC Four documentary *The Drake Equation – The Search for Life*, which examined the history and science behind this profound and enduring question.

MATTHEW COBB is Professor of Zoology at the University of Manchester, where he studies the sense of smell in maggots and teaches evolutionary biology. He is an award-winning author, teacher and translator who is currently writing a book about the history of the brain. His most recent book, *Life's Greatest Secret: The Race to Crack the Genetic Code*, was short-listed for the Royal Society Winton Book Prize.

Lewis Dartnell (www.lewisdartnell.com) is an astrobiology researcher at the University of Westminster. He studies how microbial life, and signs of its existence, might persist on the surface of Mars exposed to the bombardment of cosmic radiation, and how we could detect them. Lewis features regularly on television and radio talking about science, and his past books have included *Life in the Universe: A Beginner's Guide* and *The Knowledge: How to Rebuild our World from Scratch* (www.the-knowledge.org), a *Sunday Times* Book of the Year.

Paul Davies is a theoretical physicist, cosmologist, astrobiologist and best-selling author. He is Regents' Professor at Arizona State University, where he is Director of the Beyond Center for Fundamental Concepts in Science. He previously held academic appointments in physics, mathematics and astronomy in the UK and Australia. He has made important contributions to the theory of black holes, the origin of the universe and the origin of life. He is a Member of the Order of Australia and the recipient of numerous scientific awards including the Templeton Prize. His latest book is *The Eerie Silence: Are We Alone in the Universe?*

Chris French is the Head of the Anomalistic Psychology Research Unit in the Psychology Department at Goldsmiths, University of London. He is a Fellow of the British Psychological Society and of the Committee for Skeptical Inquiry and a Patron of the British Humanist Association. He has published over 130 articles and chapters covering a wide range of topics. His main current area of research is the psychology of paranormal beliefs and anomalous experiences. He frequently appears on radio and television casting a sceptical eye over paranormal claims. His most recent book is *Anomalistic Psychology: Exploring Paranormal Belief and Experience*.

Monica Grady is Professor of Planetary and Space Sciences in the Department of Physical Sciences at the Open University (OU)

in Milton Keynes. She has led major research programmes in the study of meteorites: her research interests are in the fields of carbon and nitrogen, and one of her major areas has been in trying to understand the history of carbon and water on Mars. In honour of her contributions to the field, the International Astronomical Union named Asteroid (4731) as 'Monicagrady'. In June 2012, she was appointed a Commander in the Order of the British Empire (CBE) for her services to Space Sciences.

NICK LANE is an evolutionary biochemist at UCL, working on how energy shapes evolution from the origin of life to the evolution of complex cells. He has published four celebrated books, translated into twenty-five languages. *Life Ascending* won the Royal Society Prize for Science Books in 2010, while Bill Gates praised *The Vital Question* as 'an amazing inquiry into the origins of life'. Nick's work was recognised by the 2015 Biochemical Society Award for his outstanding contribution to the molecular biosciences and the 2016 Royal Society Faraday Prize, the UK's premier award for excellence in communicating science.

JOHNJOE MCFADDEN is Professor of Molecular Genetics at the University of Surrey. His principal research area is investigating genetics of microbes that cause infectious diseases. He has published more than 100 articles in scientific journals on subjects as wide-ranging as bacterial genetics, tuberculosis, idiopathic diseases and computer modelling of evolution. He wrote the book *Quantum Evolution* in 2000, co-edited *Human Nature: Fact and Fiction* in 2006; and is co-author with Jim Al-Khalili of *Life on the Edge: The Coming Age of Quantum Biology* published in 2014. He is currently working on a book on Ockham's razor.

CHRIS MCKAY is a research scientist with the NASA Ames Research Center. His research focuses on planetary science and the origin of life. He is also actively involved in planning for future Mars missions including human exploration. Chris been involved in research in Mars-like environments on Earth, travelling to the Antarctic dry valleys, Siberia, the Canadian Arctic, the Atacama, Namib and Sahara deserts to study life in these Mars-like environments. He was a co-investigator on the Huygens probe to Saturn's moon Titan in 2005, the Mars Phoenix lander mission in 2008, and the Mars Science Laboratory mission, launched in 2012.

LOUISA PRESTON is a UK Space Agency Aurora Research Fellow in Astrobiology at Birkbeck, University of London. She has worked on projects for NASA and the Canadian, European and UK Space Agencies studying environments across the Earth, where life is able to survive our planet's most extreme conditions, using them as blueprints for possible extra-terrestrial life forms and habitats. She is an avid science communicator having spoken about the search for life on Mars at the TED Conference in 2013, and her first book *Goldilocks and the Water Bears: The Search for Life in the Universe* is out now by Bloomsbury Sigma. Follow her on Twitter @LouisaJPreston.

MARTIN REES is a cosmologist and space scientist. He has contributed insights to many aspects of stars, black holes, galaxy evolution, the 'big bang' and the multiverse. He is based in Cambridge, where he has been Director of the Institute of Astronomy, a Research Professor, and Master of Trinity College. He was President of the Royal Society during 2005–2010. In 2005 he was appointed to the House of Lords, and has a special interest in the risks from future technologies. He has received many international awards for his research, and belongs to numerous foreign academies. His books for general readership include *Before the Beginning, Our Final Century? Just Six Numbers, Our Cosmic Habitat, Gravity's Fatal Attraction* and (forthcoming) *What we still don't know*.

ADAM RUTHERFORD is a geneticist, writer and broadcaster. He presents BBC Radio 4's flagship science programme *Inside Science*, and a host of others on television and radio. He's also worked as scientific consultant on a number of films, including *World War Z* (2013), *Kingsman: The Secret Service* (2014), *Björk: Biophilia Live* (2014), *Life* (2017), Alex Garland's Oscar-winning *Ex Machina* (2015), and Garland's latest, starring Natalie Portman, Oscar Isaac and an alien like you've never seen before, *Annihilation* (2017).

SARA SEAGER is a planetary scientist and astrophysicist. She has been a pioneer in the vast and unknown world of exoplanets, planets that orbit stars other than the sun. Her ground-breaking research ranges from the detection of exoplanet atmospheres to innovative theories about life on other worlds to development of novel space mission concepts. Now, dubbed an 'astronomical Indiana Jones', she is on a quest after the field's holy grail, the discovery of a true Earth twin. Dr Seager earned her PhD from Harvard University and is now the Class of 1941 Professor of Planetary Science and Professor of Physics at the Massachusetts Institute of Technology. In 2015 she was elected to the US National Academy of Sciences, is a 2013 MacArthur Fellow, and was named in *Time* Magazine's 25 Most Influential in Space in 2012.

PROFESSOR ANDREA SELLA is on a mission to get us to understand chemistry and the often hidden impact it has on our lives. He does this through live demonstration lectures and by contributing to a wide range of TV and radio programmes. He is professor of inorganic chemistry at University College London, working primarily on materials synthesis and increasingly, Citizen Science. By birth Italian, he was educated in Kenya, Canada and the UK. He rides a bike, doesn't drive, and seldom flies.

ANIL SETH is Professor of Cognitive and Computational Neuro-science at the University of Sussex and Founding Co-Director of the Sackler Centre for Consciousness Science. He has published more than 100 research papers (including a couple on octopuses), and he is Editor-in-Chief of the academic journal *Neuroscience of Consciousness* (Oxford University Press). His previous books include *30 Second Brain* (Ivy Press, 2014), and *Eye Benders* (Ivy Kids, 2013; winner of the Royal Society Young People's Book Prize 2014). He lives by the sea in Brighton, and online at www.anilseth.com and @anilkseth.

SETH SHOSTAK is the Senior Astronomer at the SETI Institute, and holds degrees from Princeton University and the California Institute of Technology. In addition to professional publications, he has penned over 500 popular articles on astronomy, technology, film and television. He chaired the International Academy of Astronautics' SETI Permanent Study Group for a decade, and each week hosts the SETI Institute's hour-long, science radio show, *Big Picture Science*. Shostak has written, edited and contributed to a half dozen books, including a textbook on astrobiology and *Confessions of an Alien Hunter: A Scientist's Search for Extraterrestrial Intelligence*.

IAN STEWART is a mathematics professor at the University of Warwick and a Fellow of the Royal Society. He has published over 100 books including *Evolving the Alien* with Jack Cohen, *Professor Stewart's Cabinet of Mathematical Curiosities*, *Seventeen Equations that Changed the World* and the bestselling *Science of Discworld* series with Terry Pratchett and Jack Cohen. He has published four science fiction novels including *The Living Labyrinth* with Tim Poston. His awards include the Royal Society's Faraday Medal, the IMA Gold Medal, the AAAS Public Understanding of Science Award, the LMS/IMA Zeeman Medal, and the Lewis Thomas Prize.

GIOVANNA TINETTI is Professor of Astrophysics at University College London, where she has coordinated a team working on exoplanets since 2007. She is a Research Fellow of the Royal Society and in 2011 was awarded the Institute of Physics Moseley medal for pioneering use of IR transmission spectroscopy for molecular detection in exoplanet atmospheres.